Digital Circuit Testing

Digital Circuit Testing
A Guide to DFT and Other Techniques

Francis C. Wang

Department of Electrical Engineering
Seattle University
Seattle, Washington

Academic Press, Inc.
Harcourt Brace Jovanovich, Publishers
San Diego   New York   Boston
London   Sydney   Tokyo   Toronto

Copyright © 1991 by ACADEMIC PRESS, INC.

All Rights Reserved.

Academic Press, Inc.
San Diego, California 92101

United Kingdom Edition published by
Academic Press Limited
24–28 Oval Road, London NW1 7DX

Library of Congress Cataloging-in-Publication Data

Wang, Francis C.
    Digital circuit testing : a guide to DFT and other techniques /
Francis C. Wang.
        p.   cm.
    Includes bibliographical references and index.
    ISBN 0-12-734580-9
    1. Digital integrated circuits--Testing.   I. Title.
TK7874.W363   1991
621.381'5--dc20   90-29033
                                                                CIP

PRINTED IN THE UNITED STATES OF AMERICA
91 92 93 94   9 8 7 6 5 4 3 2 1

Contents

7. Special Testing Topics and Conclusions

Preface

Digital Circuit Testing: A Guide to DFT and Other Techniques provides an introduction to straightforward testing techniques for professional engineers involved in the design, test, and manufacture of digital electronic circuits or systems. Its content should be easily applicable to their day-to-day work.

The advent of VLSI technology in the 1980s has led to tremendous progress in the electronic industry and its product offering. In particular, digital circuits have become increasingly complex, with more transistors and functionalities packed on a single chip of nearly the same physical size. With the limited number of I/O pins on today's circuits on which to apply test stimuli and to detect test results, testing has become a major problem technically and economically. It has proven to be the weak link in the product development cycle. To compound the problem, design for testability (DFT) as a disciplined approach to solve the testing problems has not been widely accepted by the design community. One of the reasons for this is the existence of a gap between the theories involved with DFT and its applications.

This book therefore emphasizes DFT and other techniques necessary to make complex digital circuits testable. The materials are presented in an easy-to-understand manner. Important topics covered include built-in self test (BIST), scan-path and boundary scan design, mixed-signal testing and simulation, design and test of PLDs, automatic test vector generation (ATVG), and linking design with test.

For design engineers, this book should explain DFT techniques essential to the design of testable circuits so that the fundamental problems of testing can be solved at their origin. For test engineers, the book should also expose the testability limitations inherent in a design so that both design and test engineers can work together to improve the testability of a given product.

The book is organized into seven chapters. Chapter 1 presents the fundamentals of digital circuits and associated terminologies. This is followed by faulty behaviors of digital circuits and the fault models used to describe these behaviors. We then discuss the fault coverage concept and its requirements of testing digital circuits. Some fundamental ATVG algorithms are reviewed. The design and test process is then described in terms of its subprocesses and their interrelationship, with an emphasis on the role of computer-aided engineering (CAE) and computer-aided-testing (CAT) tools.

Chapter 2 is devoted to a new method of ATVG for combinational circuits using a testability measure algorithm called test counting. Practical examples are also presented. Readers uninterested in combinational circuit test generation can skip Chapter 2 and go directly to Chapter 3, which covers sequential circuit ATVG and DFT.

Since ATVG is frequently considered the bottleneck in testing, this topic is thoroughly examined. In Chapter 3, new and more efficient test generation algorithms are explored. Some practical design-for-testability techniques are also addressed. Readers will be able to see the cause and effect relationship between test generation and DFT.

Chapter 4 covers the design and test of programmable logic devices (PLDs). PLDs have gained ever-increasing applications in digital circuits/systems. Special considerations in testing PLDs, such as generating tests for non-used fuses and special algorithms derived for ATVG to take advantage of PLDs' special architecture, are discussed. We also provide extensive examples to illustrate DFT techniques for PLDs.

Chapter 5 focuses on the subjects of BIST and the newly standardized boundary scan technique for high-level assembly testing. Because test vector generation and test results verification are accomplished by built-in circuitry, BIST is becoming more important in testing high speed and complex digital circuits. Done at the circuit's operational speed, the testing does not require expensive automatic test equipment (ATE). A particularly popular BIST scheme that uses linear feedback shift registers to generate pseudorandom vectors as test stimuli and to perform signature analysis to verify test results is covered in detail. Guidelines are given to allow design engineers to implement the BIST method to achieve hardware efficiency and improve levels of controllability, observability, and testability. Chapter 5 discusses boundary scan as an emerging DFT technique for testing boards and higher level assembly. Results of recent applications of boundary scan techniques are also presented.

In Chapter 6, ATEs and their capability to perform different kinds of testing are discussed. An overview of modern ATEs for component and board level testing is first introduced, followed by a detailed discussion of the type of tests, such as in-circuit, functional, and AC and DC parametric testing. We also address the interesting topic of linking designs done with CAE tools with the testing process so that simulation results can be used in the testing environment.

In the final chapter, special testing topics, such as mixed-signal testing, are presented. Since current fabrication technology allows both analog and digital devices on a single chip, this presents additional testing problems. Another topic that deserves special attention is testing of today's 32-bit high speed microprocessors such as the reduced instruction set computers (RISC). We discuss their special testing requirements presented by the unique architecture and single chip implementation.

Although this book is written primarily for professional test and design engineers, it could also be used as either a textbook or a reference in a testing course for engineering or computer science students at the senior or graduate level.

I wish to express my gratitude for the expert assistance in preparing the book by the Academic Press staff. I also would like to thank Alan Bredon, W. T. Cheng, Song Tzer Lim, Hank Graham, Robin Cameron, Denny Siu, and Tim Kennedy for taking their time to review the manuscript. Several companies have also generously permitted me to adapt information from their publications about their products: Logic Automation Incorporated, Integrated Measurement Systems Incorporated, and Data I/O Corporation. Again, I want to express my deep appreciation for their generosity.

Finally, I would like to thank my wife Diana and my children Rossana, Geoffrey, and Rebecca for their patience and understanding during my preparation of the manuscript.

Francis C. Wang

Digital Circuit Testing

Chapter 1

Introduction

This chapter introduces the background material for the rest of this book. It begins in Section 1.1 with a brief definition of a digital circuit/system, followed by a review of some of its components at the digital logic level. Sections 1.2 and 1.3 present concepts important from a testing point of view, including fault modeling of various gates, fanout, and wired-OR. We also present an analysis of fault coverage and its requirements in digital circuit testing. Section 1.4 provides an overview of the design and test processes for today's digital circuits/systems. The importance of design automation is also summarized, with special emphasis on the role of computer-aided engineering (CAE) tools in design and test of complex digital circuits/systems. In the last two sections of this chapter we will review some important automatic test vector generation (ATVG) algorithms used in industry.

1.1 Digital Circuits and Systems

1.1.1 Definition

A digital circuit/system can be defined generally as an interconnection of logic elements such as AND gates, OR gates, INVERTERs, flip-flops, and registers. It must also be able to process a set of discrete and finite-valued electrical signals. This is in contrast to an analog circuit which is made up of electrical components such as resistors, capacitors, and transistors and which processes electrical signals of continuous values either in the form of current or voltage. An operational amplifier is an analog circuit and an ordinary oscilloscope is an analog system. A binary counter is a digital circuit and typical digital systems are digital meters, watches, and computers. The most popular digital system today is a microprocessor-based personal digital computer. In this chapter, the modern digital computer is used frequently to represent a typical digital system.

A digital computer can be viewed and studied in terms of a series of levels, one on top of the other.[1] Figure 1.1 shows a five-level digital computer. Each level treats the level below it as a virtual machine and has a set of instructions for the lower level to execute. Each level has its own interfaces to the external world or the users. It can hide all the details of the computer at other levels, and therefore it is a particular way of representing the computer with its own objectives and operations. For instance, the top level is frequently called the high-level language level or problem-oriented language level. Its objective is to provide an environment and the necessary operations for application programmers to write application programs efficiently without having to worry about how the computer works or interprets the programs.

This leveled view is one way to look at the organization of a digital computer, but it is useful because it allows engineers or computer scientists to study and work on the problems from a position much more focused on the areas of their particular expertise or interest. But of course someone will have to worry about how these different parts all fit together; this is a job for the computer architect.

At the bottom of the multilevel organization of a digital computer is the digital logic level. The digital logic level is made up of various combinational logic

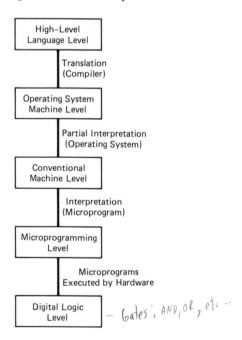

Figure 1.1 Multilevel view of a digital computer.

elements called gates and sequential logic elements called flip-flops. These elements are building blocks of the digital computer hardware since other larger blocks can be built up from them, and therefore they are frequently called primitive elements. Each element accepts a finite set of discrete values as inputs and produces a corresponding output of discrete values. These inputs and outputs normally take on the values of 1 or 0. But sometimes these values can not be determined with certainty to be a 1 or a 0. In this case, an X is used to indicate the value is unknown. Since this is the level that makes up the computer hardware it is extremely important from a testing point of view.

1.1.2 Digital Logic: Gates

Gates are combinational logic elements. They are so-called because their logic behaviors are independent of the past history of inputs and depend only on the present input. Gates are made up of physical devices called transistors. A gate is a convenient logic abstraction of these physical devices because a gate performs a well defined logic function on its input values and produces an output value. These logic functions can be represented easily by a Boolean equation. Therefore gates are used widely by both design and test engineers who work with digital circuits. It is sometimes important to understand the workings of the underlying transistors in order to understand testing-related problems such as reconvergent fanout and hazard and race conditions. Therefore we will provide a typical underlying complementary metal oxide silicon (CMOS) transistor representation for the gates defined in this chapter.

Although all gates defined in this chapter are for two-input cases, the extension to gates with more than two inputs is straightforward. In the following discussion of gates, A and B are used to represent input signals of the gate and each signal can take on binary values. The output signal is represented by Y and it takes on only binary values. In this section we limit our discussions of logic gates to the NAND, NOR, and TRISTATE gates only, because we assume that most of the readers are already familiar with the theory of digital logic. For those who are not, a good book for reference is Shiva.[2]

Figure 1.2 shows a two-input NAND gate and its transistor equivalent circuit.[3] Note that the top two transistors are *n*-type and the lower ones are *p*-type. The logic behavior of the NAND gate is shown in Table 1.1. Note that the 0 input value dominates the other input value in the sense that once a 0 exists at one input, the output automatically takes on the value of a 0. The X value is useful from a testing point of view because at times there is uncertainty with respect to the value of a particular signal, especially when a circuit is first powered up. The X is therefore used to represent the unknown values in a circuit. The NAND gate can also be considered as a series combination of an AND gate and an INVERTER.

The NOR gate is another important logic element. Figure 1.3 shows a two-

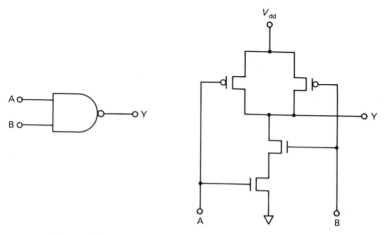

Figure 1.2 NAND gate: Symbol and its CMOS transistor circuit.

Table 1.1

Logic Behavior of NAND Gate

A	B	Y
0	0	1
0	1	1
0	X	1
1	0	1
1	1	0
1	X	X
X	0	1
X	1	X
X	X	X

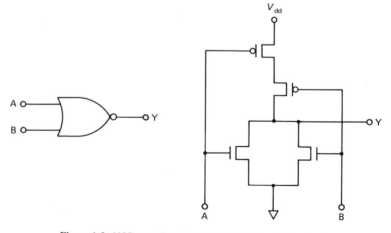

Figure 1.3 NOR gate: Symbol and its CMOS transistor circuit.

Table 1.2

Logic Behavior of NOR Gate

A	B	Y
0	0	1
0	1	0
0	X	X
1	0	0
1	1	0
1	X	0
X	0	X
X	1	0
X	X	X

input NOR gate and its equivalent CMOS transistor circuit. For a NOR gate or an OR gate the dominating input value is 1 instead of 0, as in the case of a NAND gate or an AND gate. In a similar manner to the NAND gate, the NOR gate can be considered as an OR gate connected in series with an INVERTER. Table 1.2 shows its logic behavior.

The TRISTATE (TRI) gate derives its name from the fact that it provides, in addition to a logic 1 or 0, at the output a third value called Z or high impedance. As shown in Figure 1.4, it has a single data input (i.e., A) and an enable input (i.e., B), and the output takes on the Z value whenever B is a 0 (it is disabled) as indicated in Table 1.3. When the gate is enabled (i.e., B is 1) the output signal always assumes the input signal value. However, Z is different from X because it is a value dependent on the technology of underlying transistors. A Z on the output indicates a condition in which the output is undriven. This allows other signals to drive it in a bus connection as shown in Figure 1.5a without causing conflict in signal values. The TRI gate is also frequently connected to an I/O

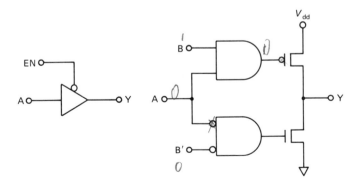

Figure 1.4 TRI gate: Symbol and its CMOS transistor circuit.

Table 1.3

Logic Behavior of TRI Gate

A	B	Y
0	1	0
1	1	1
X	1	X
0	0	Z
1	0	Z
X	0	Z

(i.e., input and output) pin in a programmable logic device (PLD) so that an input signal can be applied to the same pin when the TRI gate is disabled (i.e., with a Z at the gate output) as shown in Figure 1.5b.

1.1.3 Digital Logic: Flip-Flops

Flip-flops are sequential logic elements because their logic behavior depends not only on their present inputs but also on the past sequence of inputs. A flip-flop is used as a memory device because it can store a piece of data of logic value 1 or 0. We will discuss flip-flops in terms of their logic behaviors, gate level circuits, and most important their timing behaviors.

A flip-flop can assume two stable states of 0 or 1 at its output. Therefore it can store a 0 or a 1. We will limit our discussion of flip-flops to the D type because it is the most common flip-flop used in digital circuits. For other types of flip-flops the readers are referred to the logic design textbooks by Shiva[2] and Roth.[4]

Figure 1.6 shows the symbol of a D-type (delay or data) flip-flop. It has a single data input and a clock input C. The Q and Q' output are complements of each other. Table 1.4 shows its logic behavior. Basically, the present value on D input is sampled and gets stored in the flip-flop. In other words, the next state of the flip-flop is always the present value of the data input regardless of the present state. The D-type flip-flop is very popular because it is a simple one-bit memory.

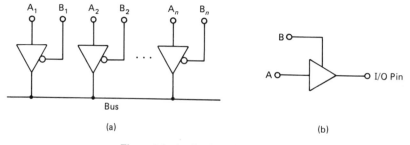

(a)

(b)

Figure 1.5 Applications of a TRI gate.

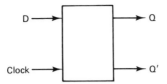

Figure 1.6   A D-type flip-flop.

Table 1.4

D Flip-Flop Logic Behavior

Q(*t*)	D	Q(*t* + 1)
0	0	0
0	1	1
1	0	0
1	1	1

It can easily latch and store any binary value coming from another digital circuit by using a clock pulse on the clock input. The stored binary value is always available at the Q output and its complement at the Q' output.

A very important property of a clocked flip-flop is the timing relationship between the input signal, the clock pulse, and the output signal. The timing diagram in Figure 1.7 shows that the output Q of the D flip-flop follows the input value starting at the transition of the clock pulse from a 0 to a 1. Because the output changes after the rising edge of the clock signal, this type of D flip-flop is also called a positive edge triggered flip-flop. It can also be designed to be negative edge triggered where the output Q follows the input D at the transition of the clock from a 1 to 0, or after the falling edge of the clock.

In order to have a flip-flop to store data correctly, the data signal must be stabilized before the clock transition occurs either from a 0 to 1 in a positive edge triggered flip-flop or from a 1 to a 0 in a negative edge triggered flip-flop. The data signal must also stay stable for a minimum time right after the clock

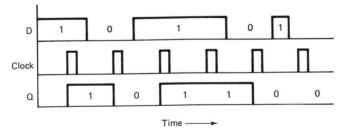

Time ⟶

Figure 1.7   Timing diagram for a D flip-flop.

transition. Now we define two terminologies related to the minimum time interval that the data signal must remain stable either preceding or succeeding the falling or rising edge of the clock.

Setup Time: The minimum time interval during which the data input signal for a flip-flop must be stable (i.e., without changing its value) before the edge of the clock pulse that triggers the latching of the data correctly by the flip-flop.

Hold Time: The minimum time interval during which the data input signal for a flip-flop must remain stable (i.e., without changing its value) right after the edge of the clock pulse that triggers the latching of the data correctly by the flip-flop.

Timing analysis is important in certain kinds of parametric testing since it is only by detecting timing violations that we can assure the correct operation of the sequential logic elements. More important is to design such flip-flops to satisfy the required timing constraints before the circuit reaches the manufacturing testing stage.

1.2 Fault Models Used in Digital Circuits Testing

We have discussed the logic behaviors of various combinational and sequential logic elements called gates and flip-flops that are the basic building blocks of any digital circuit/system. What we have not discussed is the behavior of these logic elements when some kind of failure conditions are induced by faults in the circuit. The primary purpose of testing digital circuits/systems is to find out what the faulty behaviors of logic elements are, just as a doctor must first find out the symptoms of a patient in order to prescribe a cure for the illness. In this section, we will discuss a generalized fault model independent of the process technology used in the circuit/system.

1.2.1 Definitions

In this section, we will define several terminologies related to fault models used in testing. Note that so far we have not defined testing as it relates to digital circuits/systems. We define the term *testing* here from a particular perspective, i.e., from its purpose. We will postpone other definitions of testing until we gain more knowledge about testing-related subjects. At the present, we will define several terminologies so that we can gradually get into the topics of testing and DFT techniques.

Testing: A process of evaluating a circuit/system to detect the presence of hardware failure due to faults and also to locate such faults to facilitate repair activities.

In general, testing frequently involves the application of a sequence of test stimuli called vectors or patterns to the inputs of a circuit under test (CUT) and analysis of the corresponding responses to the applied test by first collecting data at the outputs of the CUT. The analysis step is characterized by comparing the test responses with expected responses when the same test stimuli are applied to the model of the CUT. If the responses are not as expected, then it is said that a fault has been detected. The whole process is generally automated since it may involve an automatic test vector generation (ATVG) and automatic test equipment (ATE). The generation of test stimuli with expected responses is a very difficult process if done manually. ATVG can help in many cases if the design has inherent testability. The ATE used in most cases is either a components or board tester for manufacturing testing or a logic analyzer used commonly for prototype debugging.

Fault: A physical failure or defect of one or more components in a digital circuit/system caused by the manufacturing process, extreme operating conditions, or wear-out (aging) of the physical components.

Note that, from the definition above, a fault is a representation or manifestation of physical defects and as such it is a convenient term to use in the testing community. Manufacturing processes can frequently cause physical failures because of silicon defects, lithographic problems, processing problems, etc. Wear-out or aging can also cause physical defects from long-term operation of circuits under the conditions of high current densities, corrosion, ion migration, hot electronic trapping, etc. Another class of physical failures is due to the automated manufacturing steps in mounting integrated circuits (ICs) on the printed circuit board (PCB). The automated IC insertion equipment can damage the input or output pins by either bending them or shorting them out.

No matter what the reasons that cause physical failures, the results are faulty devices, opens, shorts, and improper thresholds of devices. They can cause the circuit to malfunction or degrade its prespecified performance. We use *fault* as a generic term to represent the manifestation of one or more physical defects. Some of the physical defects can cause the fault to be intermittent. This kind of fault is called a soft fault since its appearance is a function of time. In contrast, hard faults are solid faults that are consistent and permanent. Hard faults, once detected by a set of stimuli, can be detected at any other time by repeating the same test. This is the type of fault that allows the fundamental framework for fault detection and fault location techniques to be formulated.

Logic Fault: The type of fault due to physical defects of one or more components that change the logic behaviors of the digital logic elements from those defined in Section 1.2.

Logic faults are faults that change the logic behaviors listed in Tables 1.1 through 1.4. When a logic fault exists in a logic element, its normal behavior changes. For example, a metal-oxide-semiconductor field-effect transistor

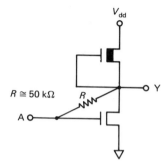

Figure 1.8 Physical defects in MOSFET INVERTER represented by logic faults.

(MOSFET) circuit normally performing the logic function of an INVERTER gate which has the physical defect of a short circuit as in Figure 1.8 changes its behavior in such a way that its output is always a 0 no matter what input values are applied. We call this effect the "output stuck-at-0" (S-A-0). If, on the other hand, there exists an open circuit between the connection of the two transistors, the output has a logic value of 1 permanently no matter what input values are applied and we say the output signal is stuck-at-1 (S-A-1). One of the tasks to test the existence of the fault is to find a set of inputs to cause the outputs to be different from the normal logic behaviors. If such a set of inputs is found, we then say this set of inputs constitutes a test for the logic element under that particular faulty condition, because the set of inputs is capable of distinguishing a logic element that is functioning normally from one that is being faulty.

1.2.2 Technology Independent Fault Model

Logic faults are technology independent in the sense that they can represent physical defects in any technology. Figure 1.9 presents an example to show how

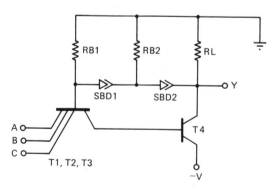

Figure 1.9 Physical defects in TTL NAND gate represented by logic faults.

Table 1.5

TTL NAND Gate Defects and Their Logic Faults

Physical defect	Logic fault
RB1 open	Output S-A-0
T1 emitter open, base open	Input A S-A-1
T2 emitter open	Input B S-A-1
T3 emitter open	Input C S-A-1
T1, T2, T3 collector open	Output S-A-0
T4 collector open, base–emitter short	Output S-A-1
T4 collector–emitter short	Output S-A-0
T1 base–emitter short	Input A S-A-1
T2 base–emitter short	Input B S-A-1
T3 base–emitter short	Input C S-A-1
T1, T2, T3, T4 collector–base short	Output S-A-0

physical defects in a transistor-transistor-logic (TTL) NAND gate can be represented as logic faults associated with the inputs and output of the gate.[5] The various defects and the corresponding logic faults are shown in Table 1.5. Hard logic faults are the primary type of faults we are concerned with in this book. We will discuss the fault models for each logic element and identify tests for the type of faults.

1.2.3 Logic Fault Models for an AND Gate

The most widely accepted fault models used for logic faults are basically the stuck-at-1 (S-A-1) and the stuck-at-0 (S-A-0) models originally proposed by Eldred.[6] In the stuck-at fault model for each gate, the inputs and the output of the logic gate are assumed to be a fixed value, either logic 1 or logic value 0, irrespective of what value is applied. Using this model, input stimuli (vectors) would be created to distinguish a good (fault-free) and a bad (faulty) model. With this type of model, test vectors can be generated for specific faults and specific defects can be associated with specific tests (vectors). The stuck-at fault model has been used for many years as a standard to generate tests for digital circuits.

Consider the logic fault model for the AND gate as shown in Figure 1.10. A two-input AND gate has six potential stuck-at faults, three stuck at 0s and three stuck at 1s. Table 1.6 shows the faulty logic behavior of the AND gate with input A stuck at 1, with the square bracket containing the faulty value of A. Note that a square bracket is also used to show faulty logic value at the output where it is different from that for a good gate. Any set of inputs for A and B that causes the output to be different from that of the good device can be used as a test to detect

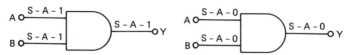

Figure 1.10 S-A-1 and S-A-0 fault model for AND gate.

Table 1.6

Faulty Logic Behavior of an AND Gate with
Input *A* S-A-1

A S-A-1	B	Y
0[1]	0	0
0[1]	1	0[1], *A* test
0[1]	*X*	0[*X*]
1[1]	0	0
1[1]	1	1
1[1]	*X*	*X*
X[1]	0	0
X[1]	1	*X*[1]
X[1]	*X*	*X*

the *A* S-A-1 fault. However, among the vectors (*A* = 0 and *B* = 1), (*A* = 0 and *B* = *X*), and (*A* = *X* and *B* = 1), only (*A* = 0 and *B* = 1) can be considered as a real fault detection test vector because the other two have *X* values at the output. As we recall from a previous section, *X* can assume either logic value 1 or logic value 0.

For the output of the same two-input AND gate stuck-at-1, the faulty logic behavior is shown in Table 1.7. It can be observed that any of the vectors

Table 1.7

Faulty Logic Behavior of an AND Gate with Output S-A-1

A	B	*Y* S-A-1
0	0	0[1], *A* test
0	1	0[1], *A* test
0	*X*	0[1], *A* test
1	0	0[1], *A* test
1	1	1[1]
1	*X*	*X*[1]
X	0	0[1], *A* test
X	1	*X*[1]
X	*X*	*X*[1]

(A = 0 and B = 0), (A = 0 and B = 1), (A = 0 and B = X), (A = 1 and B = 0), and (A = X and B = 0) can be an effective test for the output Y S-A-1 fault. It is obvious that many test vectors can detect the same fault. Also note that a single test can detect more than one fault, such as the case for the test (A = 0 and B = 1) being able to detect either input or output stuck-at-1. But this test can not distinguish the A S-A-1 fault from the output Y S-A-1 because both faults produce a 1 at the output when this vector is applied. In order to isolate the fault to a particular signal or down to the transistor level, we will need another test vector after the vector (A = 0 and B = 1) is applied. The second vector used to diagnose the fault and to locate it is (A = 0 and B = 0) since this vector will make the output a 0 if it is A S-A-1 and make the output a 1 if it is Y S-A-1.

Now we look at the output Y S-A-0 case for the two-input AND gate, and the only test as we observe in Table 1.8 is the vector (A = 1 and B = 1). Actually this is also the only test for any input stuck-at-0 faults. However, there does not exist any other vector to further isolate the fault down to the individual transistor or other physical component level. Therefore we see that the input S-A-0 fault and the output S-A-0 fault are equivalent.

Equivalent Faults: Faults that can be detected by the same test vectors and for which no vectors exist to distinguish them from one another.

1.2.4 Logic Fault Model for an OR Gate

Figure 1.11 shows a two-input OR gate with all possible logic faults identified. As noted before the logic behavior of the OR gate can be obtained from that of the AND gate by replacing 0s with 1s and vice versa. This is also true for the faulty behavior.

As noted in Table 1.9, the valid test for input A S-A-0 is the test vector (A = 1 and B = 0) because the output is definitely different from that of a fault-

Table 1.8

Faulty Logic Behavior of an AND Gate with Output S-A-0

A	B	Y S-A-0
0	0	0[0]
0	1	0[0]
0	X	0[0]
1	0	0[0]
1	1	1[0], A test
1	X	X[0]
X	0	0[0]
X	1	X[0]
X	X	X[0]

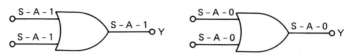

Figure 1.11 S-A-1 and S-A-0 fault models for OR gate.

free AND gate while the other two vectors ($A = 1$ and $B = X$) and ($A = X$ and $B = 0$) all involve the unknown logic value X. While in the case of the output of the OR gate stuck-at-0, any one of the five tests indicated in Table 1.10 is a valid test vector to detect this fault.

On the other hand, the output Y S-A-1 fault can only be detected by the test vector ($A = 0$ and $B = 0$) as shown in Table 1.11 and this same vector can also detect any input S-A-1 fault. However, there does not exist any vector to distinguish individual signal S-A-1 faults from each other. From a previous definition, we say that the output S-A-1 fault and the input S-A-1 faults are equivalent faults.

Table 1.9

Faulty Logic Behavior of an OR Gate with Input A S-A-0

A S-A-0	B	Y
0[0]	0	0
0[0]	1	1
0[0]	X	X
1[0]	0	1[0], A test
1[0]	1	1
1[0]	X	1[X]
X[0]	0	X[0]
X[0]	1	1
X[0]	X	X

Table 1.10

Faulty Logic Behavior of an OR Gate with Output S-A-0

A	B	Y S-A-0
0	0	0[0]
0	1	1[0], A test
0	X	X[0]
1	0	1[0], A test
1	1	1[0], A test
1	X	1[0], A test
X	0	X[0]
X	1	1[0], A test
X	X	X[0]

Table 1.11

Faulty Logic Behavior of an OR Gate with Output S-A-1

A	B	Y S-A-1
0	0	0[1], *A* test
0	1	1[1]
0	X	X[1]
1	0	1[1]
1	1	1[1]
1	X	1[1]
X	0	X[1]
X	1	1[1]
X	X	X[1]

1.2.5 Logic Fault Model for Other Gates

The INVERTER gate has a rather simple fault model because all the output S-A-1 and S-A-0 faults are the equivalent fault for the input S-A-0 and S-A-1, respectively. We will present the fault model for the output S-A-1 case in Table 1.12 and the output S-A-0 case in Table 1.13. There exists exactly one single test for each of these two cases and the single test detects the output stuck-at faults and the input stuck-at faults opposite to that of the output.

With the logic fault models described in previous sections for the AND, OR, and INVERTER gates, fault models for the NAND, NOR, and other gates can be constructed easily because each of these can be considered as a combination

Table 1.12

Faulty Logic Behavior of an INVERTER with Output S-A-1

A	Y S-A-1
0	1[1]
1	0[1], *A* test
X	X[1]

Table 1.13

Faulty Logic Behavior of an INVERTER with Output S-A-0

A	Y S-A-0
0	1[0], *A* test
1	0[0]
X	X[0]

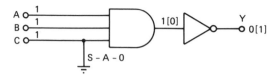

Figure 1.12 Testing of input S-A-0 for NAND gate.

of the AND, OR, and INVERTER gates. We will postpone the discussion of the
fault model for the XOR gate until Chapter 2 when we present the test generation
method using testability analysis results. The discussion of fault models related
to flip-flops will be postponed to Chapter 3 when we discuss the test generation
method for sequential circuits.

1.2.6 Other Logic Fault Model Considerations

To further illustrate the fault modeling concepts described, we now look at a
three-input NAND gate with input signal C S-A-0. The NAND gate is represented by a series combination of an AND gate and an INVERTER gate as
shown in Figure 1.12. Notice that only one test vector is required to test all three
input signals individually for S-A-0 fault. If any of the three is stuck-at-0, output
of the AND gate will always take the logic value 0; therefore, the NAND gate
will have a logic value 1 at its output because of the INVERTER gate existing
between the two outputs. All three inputs must have the logic value 1 at the same
time for testing the C S-A-0 fault. Again in this figure, we used the square
bracket to indicate the observed output when the stuck-at fault exists.

Testing of the S-A-1 fault for the same NAND gate can be done much the
same way, but in this case all inputs must be tested separately for their S-A-1
fault. We will apply a logic value 0 to each signal at the same time we apply a
logic value 1 to each of the other signals in order to detect the fault. We will
need three tests to detect all three possible input S-A-1 faults. If the output for
the NAND gate has the output of logic value 0 for any of the three tests, then we
have detected a S-A-1 fault at the input where a logic value 0 is applied in that
particular test case. For example, we assume the input A S-A-1 as shown in
Figure 1.13 and the vector applied is ($A = 0$, $B = 1$, and $C = 1$). Note that

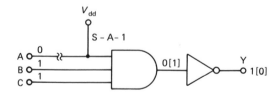

Figure 1.13 Testing of input S-A-1 for NAND gate.

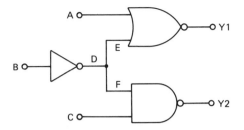

Figure 1.14 Logic fault relationship at a fanout.

the open condition at input *A* has the effect of causing this lead to have a S-A-1 fault.

The stuck-at fault model we described basically assumes that the fault is at an input or output lead (wire) of a logic gate. As shown previously in Table 1.5 a fault on the input of a logic gate may model the open-circuit condition rather than a signal stuck-at condition.[7] Therefore, a signal that fans out to more than one gate may appear normal at the input of one gate while faulty at another gate. We will illustrate this concept with an example of an INVERTER connected to a NOR gate and a NAND gate as shown in Figure 1.14. Notice that we have used different symbols (*D, E,* and *F*) to represent the output of the INVERTER, input to the NOR, and input to the NAND, respectively, because stuck-at faults on *D, E,* and *F* are not equivalent. For instance, a S-A-0 fault (due to the open circuit condition) at *E* is represented by a S-A-0 fault in Figure 1.15, but the same S-A-0 fault does not impact *D* and *F* (i.e., *D* and *F* are still normal). On the other hand, a stuck-at fault at *D* is always passed on to both *E* and *F*. The situation can become even more complicated if the stuck-at faults for *D* are propagated to *E* and *F* and later on these two signals reconverge at the inputs of another gate somewhere downstream. This is frequently called reconvergent fan-out and has tremendous impact on the efficiency of the test generation method used. However, we will postpone this topic until Chapter 2.

1.3 Fault Coverage Requirements

From the logic fault models defined in previous sections, we observe that there exist potentially $2n + 2$ stuck-at faults in an *n*-input and single-output logic gate. As logic gates are building blocks of digital circuits, a circuit potentially contains many logic gates and stuck-at faults. The total number of such faults can be in the range from thousands to hundreds of thousands. Generating test vectors to detect these faults is an important consideration in testing. From past experience, we know that to detect every potential fault that may exist in a com-

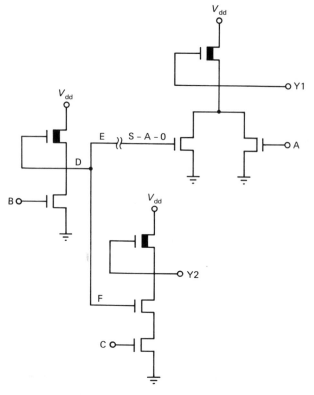

Figure 1.15 Transistor circuit for the fault relationship at a fanout.

plex digital circuit is a very difficult if not impossible task for the test engineers involved. A more realistic objective is to find test vectors to detect a high percentage of all stuck-at faults in a circuit.

1.3.1 Definition of Fault Coverage

Fault coverage is the percentage of the total number of potential faults in a digital circuit that are detected. We use the word *potential* because each fault may or may not exist in a given digital circuit. However, there is no way of knowing what faults exist in a given circuit until it is tested. In order to assure that all faults are caught if they actually exist, test vectors are needed to detect all potential faults. If this is not achievable, then guidelines must be available to test and design engineers when they are deciding on a reasonable fault coverage criterion that satisfies product-quality objectives as well as management's concern for the high costs in testing.

The product-quality objective is frequently specified in terms of the defect level based on economical factors. The cost associated with finding and replacing a defective digital circuit early in its life cycle can have an impact on the total revenue earnings of a product. As a defective circuit moves through the life cycle, the costs associated with detecting and replacing it begin to snowball.[8] To prevent marginal circuits from reaching the customer, quality assurance (QA) standards in terms of defective rates should be set stringently. Otherwise, the penalty can be substantial.

A high-quality standard can translate directly into the requirements of high fault coverage in testing. High fault coverage means more test-related resources are needed. Testing manpower and computer/tester time are both precious resources to be considered in setting fault coverage requirements. It also may require better tools, such as an ATVG that can generate the required test vectors to achieve the specified fault coverage.

Another important factor to consider is design for testability (DFT). The lack of DFT in a circuit can mean that a given fault coverage is not achievable. This in turn determines whether or not the high quality standard is obtainable. Testability is therefore both a quality and business issue. How can you ship a product to customers if you do not have enough fault coverage to ascertain its quality? In order to compete in today's market, the product's manufacturing and service costs must remain low, and this can only be done by incorporating DFT early in the design phase.

1.3.2 Fault Coverage for Circuits

The search for an appropriate fault coverage requires an analytical approach to a set of related variables. These variables include defect level, fault coverage, and yield.

Defect Level: An indicator of the percentage of circuits with defects not detected and therefore shipped with the product.

Yield: An indicator of the percentage of circuits without any faulty physical components caused by the manufacturing process.

If Y represents the yield and T the fault coverage for a given test that detects only m out of k stuck-at faults, then the defect level is[9]

$$DL = 1 - Y^{(1-T)} \tag{1}$$

Figure 1.16 shows a set of curves relating defect level versus fault coverage as a function of circuit yield.[10] If the yield is 0.25 and absolutely no testing is done (i.e., 0% fault coverage), then by the definition of yield, 25% good circuits and 75% bad circuits are shipped. In this case, test-related costs are saved, but a lot of bad products will be in customers' hands. To improve the defect level, a very high fault coverage such as 95% or above is needed in testing. To lower the

T : Fault coverage = $\dfrac{\text{detected faults}}{\text{total \# of potential faults}}$

need all fault coverage

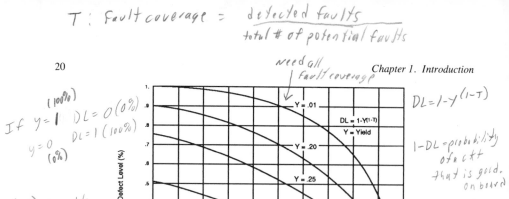

If y=1 (100%) DL = 0 (0%)
y=0 DL = 1 (100%)
(0%)

$DL = 1 - y^{(1-T)}$

$1-DL$ = probability of a ckt that is good on board

y (1-T) % ckts shipped w/ no defects

$P_C = (1-DL)^k$ = probability of all circuits are good on the board

If 2ckt, DL=.01
$P_C = (1-.01)^{20} = .82$
$P_a = 1-.82 = 18\%$: 18% chance of a board w/ fault

K = # ckts on board

need very low fault coverage

Figure 1.16 Defect level as a function of yield and fault coverage.

DL – % of ckts w/ defects not detected
Yield : % of ckts w/o any faulty physical components caused by manufact. process.

defect level to 1% or less, the fault coverage required in testing must approach 100%. For full testing (i.e., 100% fault coverage), all defects can be detected theoretically, regardless of the yield. As a result, there will not be any bad circuits shipped. This is true only if the fault model used is correct; that is, it must truly represent manufacturing defects.

How much testing is enough in terms of fault coverage for a digital circuit? With a known yield figure from the manufacturing organization and the required defect level from the QA organization, a test engineer can determine an acceptable fault coverage from the curves in Figure 1.16.

As is commonly known, fault coverage is frequently evaluated using a technique called fault simulation. Fault simulation is a CAE tool that simulates a set of test vectors to determine what faults are detected and it also calculates the fault coverage for the vectors simulated. An ATVG can also produce fault coverage figures for the test vectors it generated.

1.3.3 Circuit Defect Level and Board Failure Rate

As all digital circuits are eventually mounted on a printed circuit board (PCB) before they get shipped as part of a product, we must understand the impact of the defect level of digital circuits on the failure rates for PCBs containing many circuits. We can determine the relationship between the defect level of digital circuits, which is a function of fault coverage, and the board failure rate, assuming there are k number of circuits on a board. The defect level of these circuits is assumed to be DL. Then the probability of a circuit that is good on the boards is $1 - DL$. The probability for all circuits on the board being good is [10]

$$P_c = (1 - DL)^k \tag{2}$$

The probability for the board containing at least one bad circuit is

$$P_a = 1 - P_c \tag{3}$$

Considering a board with 20 circuits each having a defect level of 1% (i.e., each circuit has a 99% chance of being good) and using equations (2) and (3), 18% of the boards will be malfunctioning in time. It may cost a large amount of money to do the necessary warranty repair. How can a product be competitive if so much money has to be spent for maintenance and service in the field?

In summary, fault coverage must be set very high in order to reduce the defect levels of digital circuits. Even for a very low circuit defect level, the resulting board failure rates can be unacceptably high if no further testing is done beyond the circuit level.

1.4 The Design and Test Process

So far in this chapter we have discussed the building blocks of digital circuits and their fault models as well as fault coverage concepts. What we have not discussed is how these building blocks can be put together to implement a digital circuit that performs the specified functions. A prerequisite to that discussion is an understanding of the actual design and test process for complex digital circuits/systems. Digital circuit/system designers usually take a systematic approach to the design tasks involved and also use extensively varying design automation tools called computer-aided design (CAD) and computer-aided engineering (CAE).

1.4.1 Ideal Design and Test Process

In this section, we describe an ideal and simplified approach to the design and test process in terms of the phases involved (see Figure 1.17). Assuming the design of the circuit/system is based on a product requirement document, the design is started by analyzing the requirements, including its functionalities and performance specifications. Proceeding in a top-down fashion, a high-level architecture of the system in terms of its partitioned subsystems is first proposed. Functional and performance requirements for each subsystem must be specified. More importantly, subsystem interfaces must be clearly defined to allow their parallel development to reduce the overall product development time. For instance, if the product objective is to have a new digital computer to be introduced

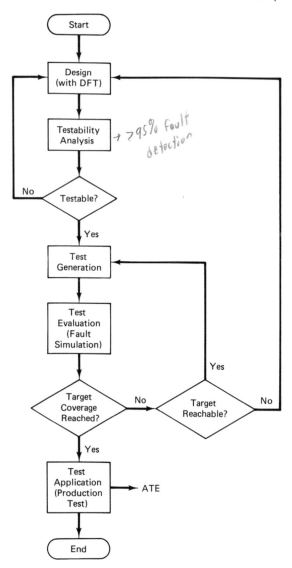

Figure 1.17   Ideal design and test development process.

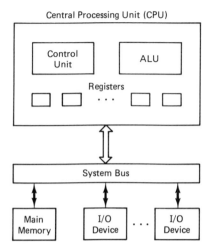

Figure 1.18   Subsystems of a digital computer.

into the market, the main units of such a computer can be the functional blocks in Figure 1.18.

Once the architecture is defined, evaluated, and agreed to by all organizations related to product development, the next step is perhaps to refine all the subsystems to functional units such as control units, arithmetic logic units, and registers in the central processing unit (CPU).

1.4.2 CAE Tools Used in Design Verification

At the functional unit level, CAE tools can facilitate the design process because the design engineers have a variety of software programs to help them to verify the functionalities and performance of each unit designed. For instance, a logic simulator using a mathematical model (e.g., netlist used to represent the connectivity of various logic elements in a digital circuit) of the unit designed and a set of design vectors can verify if the unit functions as expected. These vectors are frequently called design vectors to distinguish them from the test vectors used to detect faults caused by physical defects of circuits.

Design Vectors: Vectors intended by the design engineer to verify the correctness of the design. These vectors are frequently generated from the design engineer's understanding of the design versus fault-oriented vectors generated from knowledge of the faulty behavior of its physical defects.

For example, if the function unit is an ADDER that adds two binary numbers and produces the sum at its output, the design vectors used must verify by applying a set of two-binary numbers at the inputs and comparing the simulated output with the expected sum in binary.

If the two are identical, then the design functions correctly; otherwise the design fails to perform its intended function.

A design engineer sometimes may want to evaluate the design vectors to determine how effective they are in terms of detecting stuck-at faults by using a fault simulator. The fault simulator postulates one fault at a time and determines whether or not that fault is tested (i.e., detected). It also issues a report of the total number of tested faults versus the total number of faults and those that are untested (i.e., undetected) by the given set of vectors simulated. If a fault is untested by a given set of vectors, no matter if they are design vectors or fault-oriented vectors, it does not necessarily mean it is untestable. It only means that a set of vectors used in fault simulation was not able to detect it. But sometimes fault coverage can be low because some faults are untestable due to poor DFT.

A fault simulation can also provide information on the behavior of a digital circuit under failure conditions. For instance, it indicates the locations of faults detected and undetected in the constructed fault dictionary for a given circuit. Such information is useful for diagnosing and debugging a tested circuit. A fault simulation capability is sometimes embedded in a powerful logic simulator.

Also used widely in designing today's high speed digital circuits is a timing verifier to check critical timing relationships between signals in the circuit. It can detect violations of time constraints such as set-up time and hold time in registers and critical signal path in the circuit that violates the input-to-output time delay requirements. Some of the timing verifiers use vectors to detect timing problems and they may be part of the logic simulator tool; others use only the topology information to identify such problems. Timing verification is becoming a necessary step in design and verification of digital circuits.

1.4.3 CAE Tools Used in Testability Analysis

Testability analyzers are CAE tools used to determine at the design phase how testable a circuit would be purely from analyzing the mathematical model of a digital circuit. Tools in this category have not been fully accepted by the design community because most of the testability analysis tools do not provide results that give meaningful indications of the circuit's testability.

A useful tool must give a clear indication of the problem areas in a design in terms of testability and how to improve the design so that test vectors can be generated by an ATVG or manually to achieve high fault coverage to satisfy a corporate QA objective. A requirement for the testability analysis tool is that it must be closely related to an ATVG, because generating the necessary test vectors is the key in achieving a testable design. Test generation today is frequently the bottleneck in testing complex digital circuits.

1.4.4 Test Generation and Application

Test generation frequently involves using an ATVG to exhaustively search test vectors to achieve a high fault coverage. Inherent in an ATVG is a fault simula-

ATVG ⇌ ATPG
             ↘ pattern

tion capability to determine the fault coverage and diagnostic information for the circuit involved. Most ATVGs are based on efficient algorithms that work well for small- to medium-sized circuits that range from a few hundred to a few thousand gates. But for large and complex circuits, software coding skills and smart data structures used in the ATVG also play an important role for its efficiency in terms of computer resources needed and the program's run time.

The test vectors generated by an ATVG sometimes have to be massaged to suit other types of testing than just the pure DC functional testing for which the fault-oriented vectors are intended. Subsequently, all vectors must be ported to the ATE environment in order to run the tests on either a prototype circuit or the circuits that go into the final product. This process is called test application and it involves translation of vectors from one format to another, communications between different computers, applying input stimuli to the circuit in the tester, collection of responses from the tester, the comparison of expected responses with actual responses, analyzing data to determine problems, and producing test statistics and reports. All these steps can be automated and together with ATVG can be called collectively computer aided testing (CAT).

1.5   Review of ATVG: D-Algorithm — *slow need big computers*

In the remainder of this chapter we will review traditional ATVG algorithms and their practical limitations in testing complex digital circuits. This will shed some light on why the test generation process is considered to be the bottleneck in testing. Various DFT techniques that can alleviate the problems encounterted by today's ATVGs will be discussed in subsequent chapters.

1.5.1 A Brief History

Historically, the automatic test generation methodologies are comprised of two categories: algebraic and algorithmic. Algebraic approaches[11,12,13] generate families of test vectors as solutions to Boolean equations that describe the functions of the digital circuits involved. The major drawback of these approaches is that equations are very storage-consuming. If the equation generation is aborted in the test generation process, the solution generation may not be restarted and all the results may be lost. These drawbacks make algebraic approaches impractical for generating test vectors for large circuits.

The algorithmic approaches use various mechanisms to trace sensitive paths to propagate fault effects to primary outputs and then backtrace to primary inputs to assign logic values based on conditions set at the forward fault propagation. In this section, we review the D-algorithm[14] originated by J. P. Roth, which is the first algorithm proposed for test generation and which has been widely used for combinational circuit test generation. It has also been adapted for testing sequential circuits. We limit our historical review only to the D-algorithm be-

AND, OR, etc...   *In*   *1[0]   1[0], 1 with 1 @ c*

cause of its significance in the development of other ATVG techniques. If you are interested in these other algorithms, you are encouraged to read the book by Breuer and Friedman.[7]

1.5.2 Overview of the D-algorithm

In the past, engineers have found the terminology and notation used in the original D-algorithm complex and difficult to understand. Therefore, we use some of the terminologies commonly used in testing today to review the techniques in this algorithm.

The D-algorithm generates test vectors for one fault at a time. The key idea is to pick a fault and generate a test vector for it by sensitizing all paths forward from the location of that fault to one or more outputs to allow the fault effect to be propagated to those outputs. In the process of this forward propagation, inputs and output of gates along the sensitized paths are assigned. Once the paths are established, the process reverses its direction going back toward inputs by assigning additional input and output values to each gate until assignments of all the primary inputs of the whole circuit have been made. The values of the primary inputs constitute a test vector. This forward and backward propagation process is repeated for all faults in the circuit. The flowchart in Figure 1.19 illustrates the basic steps proposed for test generation by the D-algorithm.

Note that the step called "Find a Test" is basically to pick a test from those defined in Tables 1.6 through 1.13. The terms "Line Justification" and "Consistency Operation" were originally used by Roth. For more detailed discussion of the D-algorithm, the reader should either read the original paper by Roth or the book by Chang *et al.*[15] or Breuer and Friedman.[7]

1.5.3 Definitions of Key Terms

Key terms used in the D-algorithm and in the flowchart in Figure 1.19 are defined below. These terminologies are also commonly used to describe the procedures in the other test generation techniques.

Sensitize Paths: The technique of propagating the fault effect from its location through the input and output of other gates connected to the faulty signal until the effect reaches a primary output (PO) where it can be detected.

We illustrate this technique using a simple circuit as shown in Figure 1.20 with a S-A-1 fault at the output of the INVERTER. Since this fault can not be directly detected at the primary outputs $Y1$ and $Y2$, its effect must be propagated to a PO through either one or both paths from the faulty signal. In the D-algorithm both paths are to be sensitized and this requires a 0 assigned to A and a 1 assigned to C because we know that both E and F signals are also S-A-0 as a result of D S-A-0 (see Section 1.2.6). From Table 1.13, we also know that a 0 must be assigned to the INVERTER's input in order to detect the S-A-0 fault at its out-

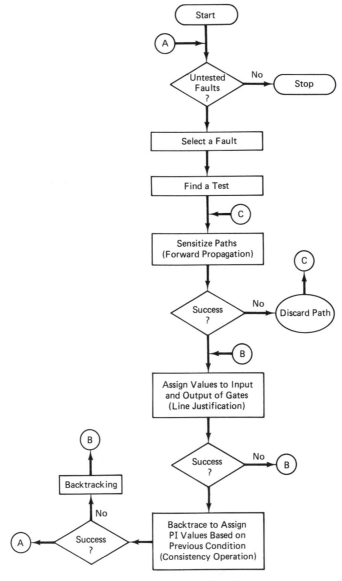

Figure 1.19   D-algorithm flowchart.

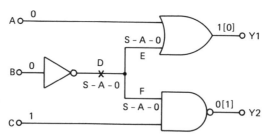

Figure 1.20 Path sensitization.

put. Therefore we have generated a test vector ($A = 0$, $B = 0$, and $C = 1$). Note that the path sensitization step is called line justification and the assignment of the primary input values is called consistency operation, as defined in the D-algorithm.

In the D-algorithm, there are two basic operations proposed to search for a test vector to test each fault. These two operations are defined below.

D Operation: The process to select a single path or multiple paths such that the sensitized value at the fault site can be driven forward until it reaches any primary output to be detected. This involves several steps, each one is called a *forward* D-drive. Starting from the fault site, each D-drive drives the sensitized value closer to the a primary output.

C Operation: The process to find a primary input test vector that will satisfy all the necessary gate input values in all sensitized paths. This process is also called backward line justification. All the input nodes with values required in a D-drive are called C-nodes. The step to justify the required value in a C-node is *backward* called C-drive. C-drive is used iteratively for all the C-nodes until the primary inputs are justified.

1.5.4 Problems with Backtracking *try a different vector*

One of the major difficulties for any ATVG using the D-algorithm is that every D-drive and C-drive involves a decision-making process, and these drives can cause conflicts in values assigned to different gates. When a conflict happens, it is necessary to return to the previous decision point and try another choice. In other words, previous conditions have to be revised (i.e., the original forward propagation process restarted and a new set of values for inputs and outputs of all the gates must be assigned). This may lead to another set of conflicts and hence this process may have to be repeated several times until no new reassignments are necessary. The original D-algorithm did not provide any suggestions for handling backtracking.

We use a simple example to show the meaning of backtracking and the harm

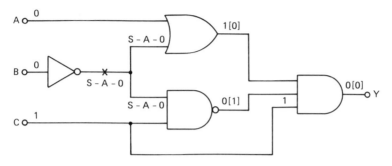

Figure 1.21 Same vector applied to the modified circuit as in figure 1.20.

it can do to the test generation process. The circuit in Figure 1.20 is modified with the two primary outputs now converging to a three-input AND gate as shown in Figure 1.21. The test vector ($A = 0$, $B = 0$, and $C = 1$) previously generated does not detect the S-A-0 fault at D any more because the fault effect propagated from both the OR and the NAND gate cancel each other at the inputs of the AND gate (i.e., the output Y of the AND gate remains at value 0 with or without this fault in the circuit). As a result, another vector must be generated to detect D S-A-0 fault. It turns out in this simple example that the new vector ($A = 1$, $B = 0$, and $C = 1$) as shown in Figure 1.22 detects this fault at the primary output Y because the fault-free output is 0 and the output with D S-A-0 is 1. In a very large circuit, backtracking can take many trial-and-error steps in order to find a set of assignments of input and output values for all the gates to allow the fault effect to be propagated to the primary outputs of the circuit.

Because of backtracking many more computational steps are needed until a set of overall consistent values can be assigned to every signal in the circuit to detect a given fault. Note that the term *backtrack* means an entirely different thing from the term *backtrace*. Backtracking should be avoided in test genera-

find input from condition needed at output

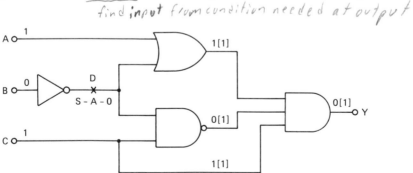

Figure 1.22 Backtracking resulted in a new vector for the same circuit as in figure 1.21.

D - all inputs need values
Podem - " " to start @ X

tion, if possible. At least the time and resources to handle it should be minimized in order to make an ATVG efficient.

more efficient

1.6 Review of ATVG: PODEM Algorithm

The classical D-algorithm is capable of generating test vectors for combinational circuits for specific faults if tests exist for these faults. However, for circuits with reconvergent fanouts commonly used as parity-checking circuits and error-correction and detection circuits, extensive backtracking may be necessary, requiring large computer resources. The path oriented decision making (PODEM) algorithm is a more efficient method for generating tests for this type of circuit because it uses heuristics to guide the backtracking search for tests to detect faults in combinational circuits.

1.6.1 An Overview of the PODEM Algorithm

The PODEM test generation method was originally developed by P. Goel.[16,17] It uses some of the notations used by the D-algorithm but the algorithm itself is different. The PODEM algorithm generates a test from the location of the fault selected and traces backward (not backtrack!) toward the primary inputs where arbitrary values are assigned intitially. Starting from the fault, an objective is set to find a test for the gate that connects to the fault and then to find inputs to the gate for the test (see Figure 1.23). In this circuit, A and B are primary inputs (PIs) and Y is the primary output (PO) with a S-A-1 fault. The first objective is to have a 0 at Y for the fault-free circuit and a 1 if the fault is present. We use the square bracket notation to show this relationship. Note that this notation is similar to the one used in the D-drive of the D-algorithm and the PODEM algorithm. To achieve this objective, a set of inputs to the AND gate must be determined that satisifies this objective; this turns out to be ($A = 1$ and $B = 1$).

Repeating this process for each gate until a primary input is reached, this input is then assigned the needed value (the rest of the PIs are still at Xs). To understand if the fault is actually detected at a PO with this new assignment of PI

primary inputs *primary output - PO*

PI -

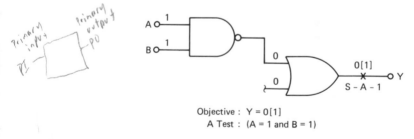

Objective : Y = 0[1]
A Test : (A = 1 and B = 1)

Figure 1.23 PODEM objective setting.

X-path - alternate P.O.

value, a simulation for both the fault-free and the faulty circuit is conducted. If it is, then the PI value establishes a sensitized path from the fault location to a PO and the task is complete. If the fault is not detected at a PO, then a forward fault propagation path (X-path) with the shortest physical length in terms of the number of gates from the fault to reach a PO is chosen. In the forward propagation, an objective is set again for each gate, just like the backtrace process. We illustrate the whole process with a flowchart in Figure 1.24.

1.6.2 A PODEM Example

A more elaborate example than the one in Figure 1.23 is given below for the circuit in Figure 1.25, in order to clarify the ideas involved with the PODEM test generation method.

1. Select a fault: output of $G2$ S-A-0.
2. Initialize all signals to Xs.
3. Set objective: $G2$ with output at 1[0].
4. Backtrace: Use Table 1.11 and Table 1.13 to set inputs to $G2$ to 0 and 0.
5. Set PI values ($B = 0$ and $C = 0$).
6. Simulate with ($A = X, B = 0, C = 0$, and $D = X$). Current fault tested at output of $G2$ (output of $G2$ has 1[0] in simulation above), but not propagated to a PO.
7. Select X-path from fault to PO Y through $G5$ and $G8$.
8. Set objective: $G5$ output at 0[1].
9. Backtrace: Set the top input to $G5$ to 0.
10. Set PI value ($A = 0$).
11. Simulate with ($A = 0, B = 0, C = 0$, and $D = X$). Current fault tested at output of $G5$, but propagated to a PO.
12. Set objective: $G8$ output at 1[0].
13. Backtrace: Note in simulation above: first input to $G8$ already set to 0, second input set to 0[1], third input to 0[X], the last input to X.
14. Set objective: $G7$ output to 0.
15. Backtrace: Set first input to $G7$ to 1 (second input already set to 0 because $C = 0$). Set second input to $G3$ to 0 (first input already set to 0 because $B = 0$).
16. Set PI value ($D = 0$).
17. Simulate with ($A = 0, B = 0, C = 0$, and $D = 0$). Current fault detected at PO Y with 1[0].

In this example we have illustrated the major steps as shown in the flowchart in Figure 1.24. After each assignment of a PI value, both the faulty circuit and the fault-free circuit should be simulated to see if the fault is tested at a particular location in the circuit and also to see if it is actually detected at a primary output.

Note that in step 4 above, the NOR gate is equivalent to an OR gate connected

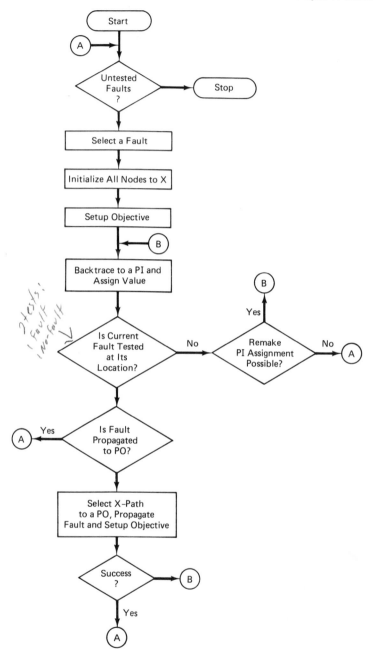

Figure 1.24 PODEM process flow.

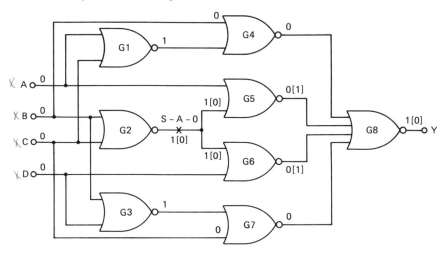

Figure 1.25 A PODEM example.

in series with an INVERTER and therefore the S-A-0 fault at the output of the NOR is equivalent to the INVERTER's output S-A-0. Since the output fault for an INVERTER is equivalent to its input stuck at the complement value, the original fault is equivalent to the INVERTER's input S-A-1. Using Table 1.11 and Table 1.13, the input to $G2$ is determined to be 0 and 0.

1.6.3 Alternative Primary Input Assignments

→ *if found conflict from previous selected values that need to be changed*

PODEM uses a method called "branch and bound" to search primary input values for tests of a selected fault.[18] In the PODEM example above, we have not encountered the situation that a PI assignment causes conflict of logic values at some other signals in the circuit that would necessitate reassignments of the original PI values. But, in reality, reassignment of the original PI value is quite common and PODEM suggests a very simple technique to facilitate the exhaustive search for all possible values.

The process uses a decision tree structure as shown in Figure 1.26 to determine the process of assigning primary input in order to generate tests. Starting with assigning all PIs to Xs, a decision node represented by a circle is reached for the initial assignment (i.e., A) of a PI. When a 1 is chosen, the left branch of the tree is taken. Otherwise the right branch is taken. Following the flow in Figure 1.24, the next PI (i.e., B) is assigned. If a branch results in no test, then the process backs up to the nearest node and makes an alternate selection of logic values (i.e., inverts the previous value). If the alternate decision also leads to a no test condition, then the node together with these dead-end branches are removed. Each time an alternate branch from a node is taken, a check mark is used

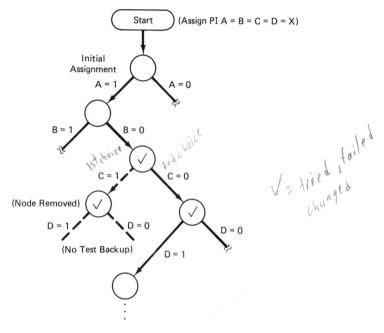

Figure 1.26 PODEM branch and bound decision tree.

to show that the other branch has been tried and discarded. This technique is useful in the value-searching process for primary input values because each occurrence of a branch being discarded results in a bounding of the decision tree and eliminates all the subsequent assignments to the as yet unassigned PI.

We illustrate the branch and bound technique by an example for the AND–OR circuit in Figure 1.27. The first choice assignment is always the left branch (i.e., PI = 1) and the resulting decision tree following the process in Figure 1.26 is shown in Figure 1.28. Note that only one PI assignment was remade in this simple circuit. Had we always first selected the right branch, then more PI reas-

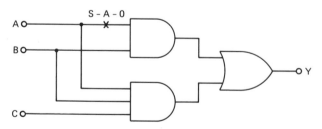

Figure 1.27 Circuit example for PODEM branch and bound technique.

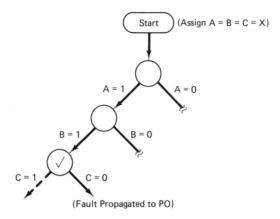

Figure 1.28   Decision tree for circuit example in figure 1.27.

signments would have been made. The reader may like to try using the right branch as the first choice at each decision node to generate vectors and see how many more PI remakes will be needed.

PODEM is just like the D-algorithm in that it is a complete algorithm. Both can generate tests to detect faults if such tests actually exist. However, PODEM is more efficient in terms of computer time used to generate the test vectors. It also uses some simple heuristics to improve its capability. Therefore, PODEM is considered a better test generation technique for complex combinational circuits.

References

1. Tanebaum, Andrew S., "Structured Computer Organization." Prentice-Hall, Englewood Cliffs, New Jersey, 1984.
2. Shiva, Sajjan G., "Introduction to Logic Design." Scott, Foresman and Company, Glenview, Illinois, 1988.
3. Weste, Neil, and Kamran Eshraghian, "Principles of CMOS VLSI Design: A System Perspective." Addison-Wesley, Reading, Massachusetts, 1985.
4. Roth, Charles H. Jr., "Fundamentals of Logic Design." West Publishing, St. Paul, Minnesota, 1985.
5. Abraham, Jacob A., "Introduction to DFT," Oregon Graduate Center—DFT Workshop, December 1984.
6. Eldred, R. D., Test routines based on symbolic logic systems, *Journal of ACM* **6**(1), 1959.
7. Breuer, Melvin A., and Arthur D. Friedman, "Diagnosis & Reliable Design of Digital Systems." Computer Science Press, New York, 1976.
8. Davis, B., "The Economics of Automatic Testing." McGraw-Hill, New York, 1982.
9. Williams, T. W., "Trends in Design for Testability." Kluwer Academic Publishers, Norwell, Massachusetts, 1988.

10. Wang, F., Denny Siu, and Tim Kennedy, Fault-coverage considerations in testing PLDs, *Evaluation Engineering*, December 1989.
11. Armstrong, D.B., On finding a nearly minimal set of fault detection tests for combinational logic nets, *IEEE Transactions on Electronic Computers* **EC-15**, February 1966.
12. Sellers, F. F., M.Y. Hsiao, and C. L. Bearson, Analyzing errors with the Boolean difference, *IEEE Transactions on Electronic Computers* **EC-17**, July 1966.
13. Akers, S. B., Universal test sets for logic networks, *Proceedings of Switching and Automata Theory Symposium*, October 1972.
14. Roth, J. P., Diagnosis of automata failures: a calculus and a method, *IBM Journal of Research and Development* **10**, July 1966.
15. Chang, Herbert Y., Eric Manning, and Gernot Metze, "Fault Diagnosis of Digital Systems." Krieger, Melbourne, Florida, 1974.
16. Goel, P., An implicit enumeration algorithm to generate tests for combinational circuits, *IEEE Transactions on Computers* **C-30**, March 1981.
17. Goel, P., and B. C. Rosales, PODEM-X: an automatic test generation system for VLSI logic structures, *Proceedings of 18th Design Automation Conference, 1981*.
18. Lawler, E. W., and D. E. Wood, Branch-and-methods—a survey, *Operations Research* **14**, 1966.

Chapter 2

Skipping until later

A Test Generation Method Using Testability Results

2.1 An Overview

Traditionally, testability measures were used to derive information on how testable a circuit was early in the design cycle so that the design could be modified to improve its testability. Testability information has also been used to aid the test generation process. We have seen in Chapter 1 that the PODEM algorithm uses heuristics to guide the test generation process in search of test vectors. Most testability algorithms [1,2,3] produce results on the basis of circuit topology descriptions, and these results are in general in the form of the circuit's controllability and observability measures. Their objective is to estimate whether further design work must be done in order to enhance the testability of the circuit.

These testability algorithms have been facing the most basic question: whether or not they provide accurate estimates of the ease with which a circuit can be tested. The SCOAP algorithm [1] in particular has been evaluated to correlate its testability measure and ease of fault detection. [4,5] A statistical approach to evaluating testability algorithms was used, basically involving comparison of testability measure results with fault coverage results for test vectors generated from the same circuit description. Evaluation results indicated only a moderate relation existing between its testability measure and fault detection. The correlation is less than 0.5. These results suggested that extreme caution should be exercised in using such measures to make design decisions and that researchers should focus on developing better testability measure algorithms.

In this chapter, we present a test generation method for combinational circuits that derives test vectors directly from testability measure results. We call it the testability–measure method of test generation. The testability information used is based on a new testability algorithm called Test Counting developed by Akers and Krishnamurthy.[6] The materials used in the next section are adapted from Ref. 6. The objective of Test Counting is to provide a meaningful indication of the scope of the testing problem for the circuit. The testability indicator Test Counting gives is an estimate of the number of test vectors (i.e., test-set size) needed to detect user specified faults. The algorithm grows linearly with respect to gate counts for combinational circuits and therefore it is very efficient. It is a quite different but a very useful approach to testability analysis because the test-set size is always a serious concern for the testing engineer. This is due to the fact that a large test-set requires more time to run on the ATE and that it may exceed the limitation of the number of vectors allowed for some old equipment.

We encourage the reader to be patient in reading through the next several sections because of the new concepts and techniques involved. The basic approach of test generation using testability measure results is shown in a flowchart in Figure 2.1. Also note that in this method of test generation there is no need to run a fault simulator to determine fault coverage for the test vectors generated.

For traditional testability measure algorithms, the actual testing requirements have seldom been considered in deriving the testability information of a circuit. The Test Counting algorithm, however, uses the actual testing requirements (i.e., the faults to be detected) to propagate sensitivity values and logic values across the circuit based on the derived relationship for these values between inputs and

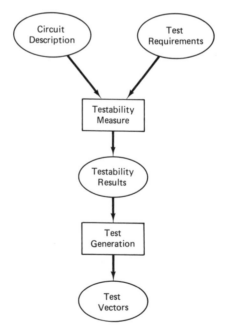

Figure 2.1 Test generation directly from testability measure results.

outputs for various types of gates. With this approach, the faults to be detected can be prespecified as input parameters to the testability analyzer. As such, the percentage of fault coverage and the number of tests (i.e., the output of the testability analyzer) to detect the specified faults are known prior to test generation.

The Test Counting algorithm can also be used to provide global testability information for each signal in the circuit in terms of the number of times that the signal takes on different sensitivity values in test generation. This result is very useful in aiding the test generation process to derive the set of test vectors to detect the prespecified faults. The test generation method described in this chapter enumerates the sensitivity values of each signal in the circuit and assigns proper logic values to it to derive test vectors[7] that can detect all the faults specified. In this chapter we first present some new notations and concepts used by Test Counting so that later we can describe the testability–measure test generation method itself.

2.2   Background in Test Counting

The testability–measure method of test generation is based on the Test Counting algorithm, which propagates user specified test requirements in terms of faults to be detected through all the gates in a circuit to estimate the size of the test-set to achieve these requirements. To present the testability–measure method, it is first necessary to describe the concept of Test Counting as it is used to determine the testability of a given circuit in terms of the number of test vectors needed to test specified faults. According to the Test Counting algorithm, "sensitivity" values for each lead are propagated in the network of the circuit to be tested using a simple algebraic relationship between input and output sensitivity values for different gates.

Sensitive Value: A logic value V on a lead A is sensitive if and only if (iff) the fault S-A-V' on the lead is observable (i.e., detectable), where V' is the complement of V.

Insensitive Value: A logic value V on a lead A is insensitive if and only if the fault S-A-V' on the lead is not observable.

Sensitive values of logic 0 and 1 are denoted by 0^+ and 1^+, respectively, and insensitive 0 and 1 are denoted by 0^- and 1^-. Referring to Figure 2.2, the sensitive (and insensitive) values are indicated on the circuit under ($A = 1$, $B = 0$, $C = 0$, and $D = 1$). The sensitive values, when complemented, are observable (i.e., detectable) on the output lead, and insensitive values, when complemented, are not observable on the output lead. For instance, the 0 input

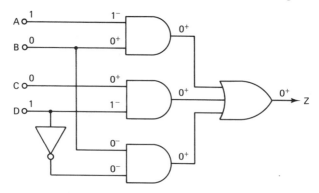

Figure 2.2 A test vector and the test values.

to the top AND gate is sensitive because if it changes to a 1 (i.e., it is complemented), then the AND gate's output changes from a 0 to a 1 and this change
can be observed (detected) at the circuit's output lead Z. On the other hand, the
1 input to this AND gate is insensitive because when it changes to a 0, the output
of the AND gate remains the same and therefore is not observable (detectable)
at the output lead Z. The more interesting case is the lower AND gate. Note here
that either input is insensitive because when it changes from a 0 to a 1 the AND
gate's output changes. However, the output of this AND gate is itself sensitive
with respect to 0 (i.e., it has a 0^+) due to the fact that if it changes from a 0 to
a 1 for reasons such as the existence of a stuck-at-1 fault on the AND gate's
output lead, this change is observable (detectable) at the output lead Z. The
reader is encouraged to verify these in order to gain an understanding of the
fundamental concepts used in the Test Counting and the testability–measure
method of test generation.

Note that for any test vector applied to a circuit, each lead assumes one of four
values: 0^+, 1^+, 0^-, and 1^-. These four values are called *test values*. During
testing (i.e., applying successive test vectors to detect specified faults), the total
numbers of sensitive 0s and 1s a lead A assumes are respectively denoted as A_0^+
and A_1^+. The total number of insensitive 0s and 1s are respectively as A_0^- and
A_1^-. These four quantities A_0^+, A_1^+, A_0^-, and A_1^- are called *test counts*. In the
following context, we will use N_0^+, N_1^+, N_0^-, and N_1^- to represent test counts for
an arbitary lead N in a circuit. A test count matrix is a 2×2 matrix with N_0^+,
N_1^+, N_0^-, and N_1^- at upper left, upper right, lower left, and lower right corners,
respectively (see Figure 2.3a), and a 3×3 test count matrix may also be used
to carry extra information about the sum of each column and row in the 2×2
test count matrix (see Figure 2.3b). In this case, the 2×2 matrix is exactly the
upper left portion of the corresponding 3×3 matrix. The numbers in the lower

N_0^+	N_1^+	$N_0^+ + N_1^+$
N_0^-	N_1^-	$N_0^- + N_1^-$
$N_0^+ + N_0^-$	$N_1^+ + N_1^-$	$(N_0^+ + N_1^+ + N_0^- + N_1^-)$ or $(N_0^+ + N_0^- + N_1^+ + N_1^-)$

N_0^+	N_1^+
N_0^-	N_1^-

(a)

(b)

Figure 2.3   A test count matrix.

right corner of the 3×3 matrices for all leads are used by Test Counting to derive an estimate of the number of test vectors needed to test the circuit.

Test counts on the input leads of a gate and test counts on the output of the same gate must satisfy certain relationships called constraints. Due to these constraints, changes of test counts on the input leads of a gate will cause changes on the output test counts. We call this forward propagation of test counts. Conversely, changes of the input test counts caused by a change of the output test counts is called backward propagation of test counts. Both forward and backward propagations are dictated by constraints to be illustrated below. Note that a constraint is nothing but an algebraic relationship between the test counts of the input and output leads of a gate.

An example of an AND gate is considered first. As illustrated in Figure 2.4, whenever the output lead C of an AND gate is required to be a 0^+, at least one of the input leads must be 0, which is sensitive iff the other input, say B, is 1^-. Since this case and the other case (i.e., the case B is 0^+ while A is 1^-) are mutually exclusive, it can be concluded that the number of times C is sensitive 0 is at least equal to or greater than the sum of times when A and B are independently being sensitive 0. The following equation expresses this constraint or relationship mathematically.

$$C_0^+ \geqslant A_0^+ + B_0^+ \tag{1}$$

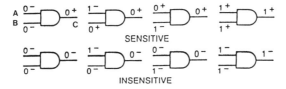

Figure 2.4   Test values for an AND gate.

On the other hand, whenever one input lead is required to be 1^+, the output and the other input leads must all be 1^+ in order to satisfy the definition of sensitive value given earlier. Hence,

$$C_1^+ \geqslant \text{Max}(A_1^+, B_1^+) \tag{2}$$

Equations (1) and (2) are the constraints for test count forward propagation through an AND gate.

Backward propagation also affects test counts. Figure 2.4 shows all eight possible cases: four for the output being sensitive and four for the output being insensitive. Note the first case is for the output being independently sensitive from the inputs (in order to detect the output S-A-1 fault when the gate is embedded in a circuit as in the case of Figure 2.2). We can also observe that whenever an input lead is 1^-, either the other input lead is 0^+ or the output lead is 1^-. Since both cases are mutually exclusive, there exists the following constraint:

$$A_1^- \geqslant B_1^+ + C_1^- \tag{3}$$

Also, an input lead assumes 1^+ whenever the output lead is 1^+ or the other input lead is 1^+. Therefore, we have the following constraint:

$$A_1^+ \geqslant \text{Max}(B_1^+, C_1^+) \tag{4}$$

We can combine equations (2) and (4) to have,

$$A_1^+ = B_1^+ = C_1^+ \tag{5}$$

The constraints for an AND gate backward (in)sensitivity value propagation are given in equations (3) and (5).

We can obtain the constraints for an OR gate by simply interchanging 0s and 1s from the above equations for an AND gate and they are listed below.

$$C_1^+ \geqslant A_1^+ + B_1^+ \tag{6}$$

$$C_0^+ \geqslant \text{Max}(A_0^+, B_0^+) \tag{7}$$

$$A_0^- \geqslant B_1^+ + C_0^- \tag{8}$$

$$A_0^+ = B_0^+ = C_0^+ \tag{9}$$

Equations (6) and (7) are constraints for an OR gate sensitivity value forward propagation and equations (8) and (9) are for its backward propagation.

For an INVERTER, the output lead is sensitive iff the input lead is sensitive with opposite logic values. Using A and C for input and output lead respectively, the constraints are given below:

$$A_0^+ = C_1^+ \tag{10}$$

$$A_1^+ = C_0^+ \tag{11}$$

$$A_0^- = C_1^- \tag{12}$$

$$A_1^- = C_0^- \tag{13}$$

The above four equations are applicable to both forward and backward sensitivity value propagation.

The constraints for a NAND gate can be derived by considering it as an AND gate followed by an INVERTER.

$$C_1^+ \geq A_0^+ + B_0^+ \tag{14}$$

$$C_0^+ \geq \text{Max}(A_1^+, B_1^+) \tag{15}$$

$$A_1^- \geq B_0^+ + C_0^- \tag{16}$$

$$A_1^+ = B_1^+ = C_0^+ \tag{17}$$

For a NAND gate, equations (14) and (15) are for forward sensitivity value propagation while equations (16) and (17) are for backward sensitivity value propagation.

Similarly, the constraints for a NOR gate can be derived by considering it as an OR gate followed by an INVERTER.

$$C_0^+ \geq A_1^+ + B_1^+ \tag{18}$$

$$C_1^+ \geq \text{Max}(A_0^+, B_0^+) \tag{19}$$

$$A_0^- \geq B_1^+ + C_1^- \tag{20}$$

$$A_0^+ = B_0^+ = C_0^+ \tag{21}$$

For a NOR gate, equations (18) and (19) are for forward sensitivity value propagation while equations (20) and (21) are for backward sensitivity value propagation.

For an XOR gate, different combinations of sensitive values of N_0^+ and N_1^+ are possible on the output lead for the same sensitive values on the input leads. Figure 2.5 shows a two-input XOR gate. By examining these different cases of test values, we see that if input A has a 0^+, input lead B and output lead C can have two possible combinations of sensitive values of (0^+ and 0^+) or (1^+ and 1^+). Note, however, that the total sensitive value ($N_0^+ + N_1^+$) remains the same in each of the different combinations and equals the maximum total sensitive values among the the input leads. Figure 2.6 shows an example where it is possible to have three combinations of sensitive values (i.e., 7¦3, 5¦5, and 3¦7) on

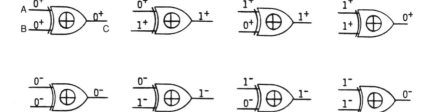

Figure 2.5 Test values for an XOR gate.

the output lead for the same input sensitive values. It is also true that an input lead is sensitive (i.e., having a 0^+ or a 1^+) iff the other input lead and the output lead are both sensitive. The output lead is sensitive iff all input leads are sensitive. Two constraints for forward and backward (in)sensitive value propagation are provided in equation 22 and equation 23:

$$A^+ = B^+ = C^+ = \text{Max}(A^+, B^+, C^+) \tag{22}$$

$$A^- = B^- = C^- = \text{Max}(A^-, B^-, C^-) \tag{23}$$

Sensitive values on a fanout node must be treated carefully because these values propagating along different branches may reconverge and cancel each other out. The sensitive values can not always be propagated through a fanout node due to the fact that different configurations associated with the various branches may impact the sensitive value propagation. A careful analysis must be performed to determine the rules for propagation. This is the subject of Section 2.3.

However, logic values must be the same on the stem and its branches when the same test vector is applied to the circuit. Therefore, two constraints are given below for the logic value propagation for the fanout circuit with stem A and two branches B and C as shown in Figure 2.7.

$$A_0 = B_0 = C_0 = \text{Max}(A_0, B_0, C_0) \tag{24}$$

$$A_1 = B_1 = C_1 = \text{Max}(A_1, B_1, C_1) \tag{25}$$

For constraints derived for the general cases of more than two inputs for various gates, the reader is encouraged to consult Akers and Krishnamurthy[6] in order

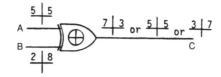

Figure 2.6 Test counts propagated through an XOR gate.

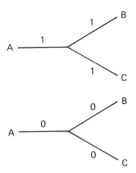

Figure 2.7 Test values on a fanout node.

to program the algorithm in software. Utilizing the constraints presented in this section, test count matrices can be generated for all the leads in a circuit through forward and backward propagation of sensitive and insensitive values. Figure 2.8 shows a circuit with a test count matrix calculated for each lead using the constraints presented in this section. Note that the estimated number of tests for this circuit is six (i.e., four sensitive 0s and two sensitive 1s for the output lead Z as indicated by its test count matrix). The calculation for the test count matrices is fast since Akers and Krishnamurthy claimed that the algorithm terminates after at most two or three passes of forward and backward propagation of (in)sensitive values through most circuits.

With this much background in Test Counting, we are ready to present the

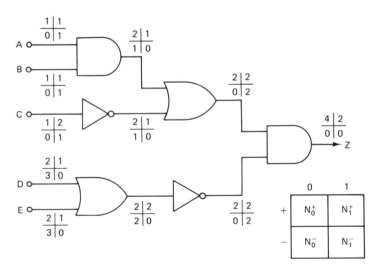

Figure 2.8 Propagation of test counts in a circuit.

testability–measure method of test generation. However, we need to address the handling of sensitive and insensitive values through reconvergent fanouts further because the resolution on sensitivities is lost when the test counts are propagated through a fanout due to potential cancellation of these values at a reconvergent node. We will present the results of an analysis of sensitivity propagation through a reconvergent fanout in the next section.

2.3 Sensitivity Analysis at a Fanout Node

In order to use the Test Counting–based testability results to generate test vectors, we extended the original Test Counting algorithm to handle propagation of sensitivity values through a fanout node. The materials in this and next section are based on the work of Hung and Wang.[7]

With respect to the fanout node in Figure 2.9, we define the following terminologies.

Fanout Stem: The lead in a fanout node where the signal originates. It is the *A* lead in Figure 2.9. For instance, the stem can be the output of an AND gate that drives other gates connected to it.

Fanout Branch: One of many leads connected to the stem. A branch must carry the signal with the same logic values as the stem.

Reconvergent Node: A gate with several inputs and some of the signals carried are originated from the same stem.

As we see in equations (24) and (25), the logic values on the branches and the stem are the same and also the logic value count equals the maximum logic value count among the stem and the branches. However, the resolutions of sensitivities are generally lost while test counts propagate through a fanout node due to the following reasons.

1. *Self masking:* The setting of sensitive value on a branch can be masked by the setting on the other branches. This can make the stem insensitive while the branches are sensitive.

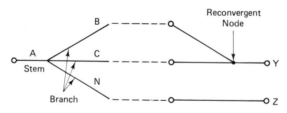

Figure 2.9 Fanout and reconvergent node.

2. *Multiple-path sensitization:* Some faults require propagation of fault effect along multiple paths for detection. This may make the stem sensitive but not the branches.

3. *Redundancy:* Some faults are not detectable because of the existence of redundant logic in the design. This causes the loss of sensitivity along the reconvergent paths.

This lost sensitivity resolution can be restored by a technique called *local enumeration* applied along the subcircuit between the fanout node and the reconvergent node. We call the subcircuit the *fanout loop.* In order to introduce the local enumeration technique, we must understand the relationship among the sensitivity values of the branches and the stem by analyzing different configurations commonly encountered in a real digital circuit. With reference to Figure 2.10, we define the following terminologies in order to present a complete analysis of the fanout problems.

D-Path: A fanout path with sensitivity on the branch and the reconvergent point. Note that the change of logic value on the branch causes change on the reconvergent point by definition of being sensitive. The symbol n_D is used to denote the total number of D-paths in a reconvergent fanout.

ND-Path: A fanout path with insensitivity on the branch and sensitivity on the reconvergent point. Note that the change of logic value on the branch does not cause change on the reconvergent point, thus an ND path is always a disabled path. The symbol n_{ND} is used to denote the total number of ND-paths in a reconvergent fanout.

U-Path: A fanout path with insensitivity on both the branch and the reconvergent point. Note that a U-path can be either an enabled path or a disabled path. The change of logic value on the branch causes change on the reconvergent point iff the path is enabled. The symbol n_U is used to denote the total number of U-paths in a reconvergent fanout.

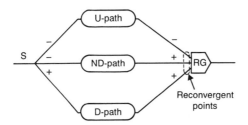

Figure 2.10 Different types of paths in a reconvergent fanout loop.

Table 2.1

Sensitivity Analysis at a Fanout Stem

				RG		XOR RG	
Case	n_D	n_{ND}	n_U	Partial	All	Partial	All
1	$=0$	$=0$	>0	S^-	S^+ iff multiple-path sensitization S^- otherwise	S^-	S^-
2	$=0$	>0	$=0$	S^-	S^-	S^-	S^-
3	$=0$	>0	>0	S^-	S^-	N.A.	N.A.
4	>0	$=0$	$=0$	S^+	S^+	S^+ iff n_D is odd S^- otherwise	S^+ iff n_D is odd S^- otherwise
5	$=1$	$=0$	>0	S^- iff some U-paths enabled S^+ otherwise	S^- iff some U-paths enabled S^+ otherwise	N.A.	N.A.
6	>0	>0	$=0$	S^+	S^+	S^- iff n_D is even S^+ otherwise	S^- iff n_D is even S^+ otherwise

Also note that for any gate except the XOR gate there are three possibilities of sensitive combinations for its inputs: none are sensitive, only one is sensitive, or all are sensitive. For an XOR gate, either all inputs are sensitive or all are insensitive. When applying the above observation to a reconvergent gate, the sensitive values of the stem S can be categorized into six possible cases as shown in Table 2.1. The symbols S^+ and S^- stand for sensitive and insensitive stems, respectively. The columns labeled "Partial" mean that the reconvergent paths of a fanout account for a subset of all the input leads of the reconvergent gate. The columns labeled "All" mean that all the input leads of the reconvergent gate are originated from the same fanout.

2.4   Local Enumeration Technique

The purpose of local enumeration is to propagate a sensitive (i.e., 0^+ or 1^+) value on a branch or an adjacent lead not belonging to the fanout path toward the reconvergent gate, meanwhile figuring out the necessary sensitive values and logic values on all other branches. More importantly, it determines whether the stem is sensitive based on the analysis in Table 2.1. By local enumeration, we shall be able to identify all the redundancy (if any) and understand the test requirement (in terms of sensitivity) propagation conditions for the fanout loop.

2.4.1 Description of the Local Enumeration Process

Flowcharts in Figures 2.11 through 2.15 illustrate in various levels of detail the local enumeration technique. Note that TMA stands for Testability–Measure Analyzer which is a general purpose digital computer software comprising known Test Counting procedures. TMA is invoked first before local enumeration to obtain the testability information in terms of test count matrices needed by the local enumeration steps in the flowcharts. Associated with local enumeration technique, certain logic redundant faults on the fanout branches can also be discovered. We will discuss the use of this technique for logic redundancy detection and elimination in Chapter 3 together with DFT considerations.

As we mentioned before, local enumeration is to provide partial solutions for sensitivity propagation through reconvergent fanout loops, and it is the first part of the overall enumeration process to generate test vectors from test count matrices for a circuit. Referring first to Figure 2.11, the process starts with an unenumerated innermost reconvergent fanout loop. It then proceeds to the block "Enumerate the propagation conditions for this fanout loop," which is further described between P1 and P2 in Figure 2.12. The "Redundancy check" block is further represented in more detail between P3 and P4 in Figure 2.13. In this figure, the first decision block is entered to ascertain whether any partial solutions have been produced. If some partial solutions have already been formulated, it means that some of the test counts in the test count matrices for leads associated with the reconvergent fanout loop have been accounted for and that therefore the partial solutions should be subtracted from all corresponding test counts. For instance, referring to the test count matrix in Figure 2.3, if a partial solution for the given lead includes a 1^+, then the sensitive 1 count in the matrix should be reduced by one, indicating this count of the test count matrix has been satisfied once. The goal in local enumeration is to satisfy as many counts in the test count matrices as possible.

The process is then directed to the decision block (in Figure 2.13) to determine whether or not all branches from the same stem have been enumerated. If all branches have been enumerated, then the decision block to determine if all adjacent leads have been enumerated is entered. (An *adjacent lead* is defined as an input lead for which a branch of the fanout node is also an input to the same gate.) If the answer is no, an enumerated adjacent lead is chosen as the driving lead for sensitivity propagation according to the rules as illustrated in Figure 2.16. A circle is used to mark the driving lead. Input leads are chosen for driving starting from the deepest branch, that is, the branch (or adjacent lead) farthest away from the reconvergent gate.

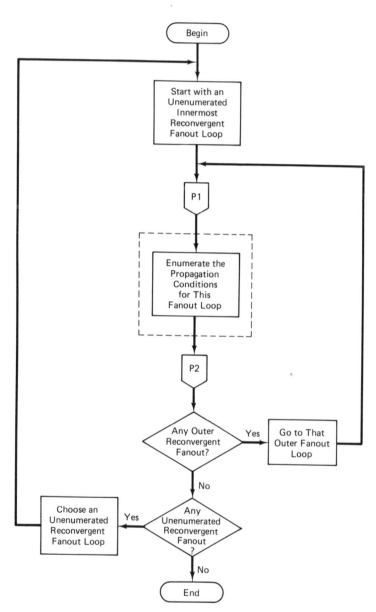

Figure 2.11 Flowchart for fanout handling: an overview. (HERE)

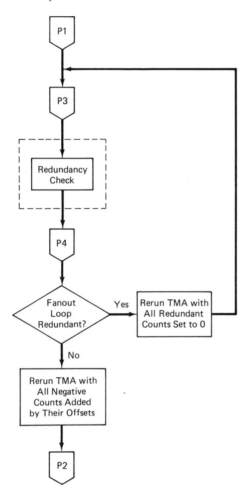

Figure 2.12 Flowchart for fanout handling: enumerate the propagation condition.

In Figure 2.13, the process proceeds to the "Loop enumeration for a sensitive value on the driving lead" block, which is further expanded to the flowchart in Figure 2.14.

Referring now to Figure 2.15, in the top block the selected driving lead is set to an enumerated sensitivity value 0^+ or 1^+. In the next decision block, it is determined whether this driving lead is a reconvergent point (i.e., at the end of a reconvergent fanout loop). If the answer is no, the forward sensitivity drive continues from gate to gate with the input lead considered as the driving lead until the reconvergent point is reached. In this block, it is determined whether the driving lead is an input to an XOR gate, an input to another type of gate, or

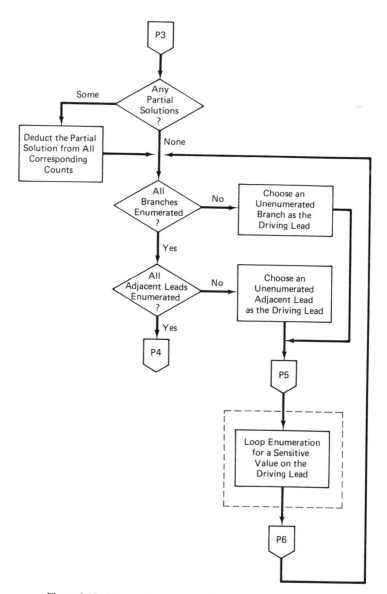

Figure 2.13 Flowchart for fanout handling: Redundancy detection.

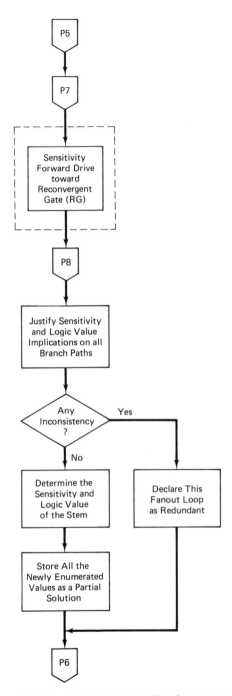

Figure 2.14 Flowchart for fanout handling: Loop enumeration.

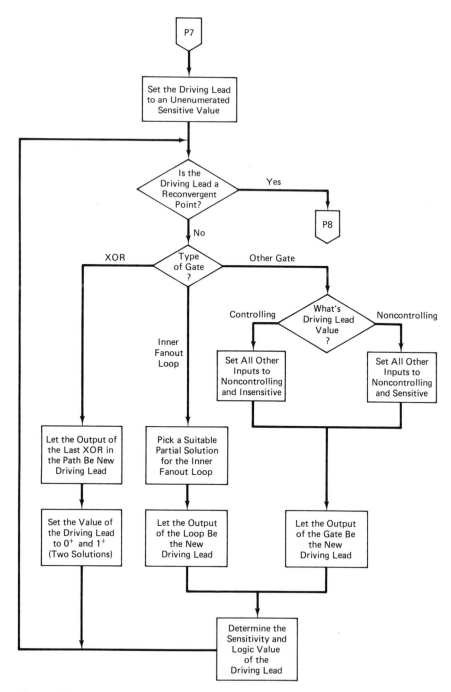

Figure 2.15 Flowchart for fanout handling: Sensitivity forward drive and reconvergent gate handling.

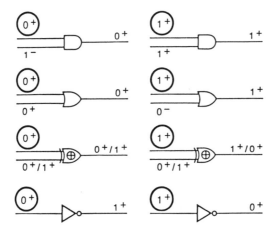

Figure 2.16   Forward sensitivity drive requirements for different gates.

an inner fanout loop. In the third instance, if the driving lead is part of an outer reconvergent fanout loop encompassing an inner loop which has already been enumerated, then a suitable partial solution which is consistent with the present solution is taken.

However, if the decision block "Type of gate?" yields another type of gate than XOR, a determination is made regarding the driving lead value (i.e., whether it is "controlling" or "noncontrolling"). A controlling driving lead value is a logic value which solely determines the output of the gate. Thus, for example, a 0^+ at the input to an AND gate is controlling when all other inputs are 1^-. A driving lead value is noncontrolling if the logic value does not solely determine the output of the gate. Thus, a 1^+ input to an AND gate is noncontrolling if all the other inputs are 1^+. The setting of all other inputs in order to allow the sensitivity value to drive through the gate to its output are based on the forward sensitivity drive requirements for various gates shown in Figure 2.16.

To carry out the local enumeration process, sensitivity and insensitivity must be driven across various gates in the fanout loop just like the forward and backward propagation of test counts. Referring to Figure 2.16, the input lead marked with a circle denotes a "driving" lead and all other leads are the "driven" leads. For instance, if the driving lead of an AND gate is required to have a 0^+, all other driven input leads must be set to 1^- and the driven output lead must have a 0^+. Figures 2.17 and 2.18 illustrate backward sensitivity and backward insensitivity drive. Note that there is no forward insensitivity drive.

Again referring to Figure 2.15, if the type of gate is an XOR gate, either output of the last XOR gate in the path (assuming more than one), or the single XOR gate if only one exists, is made the new driving lead. Two distinctive solutions are provided by choosing the value of the driving lead to be 0^+ for one

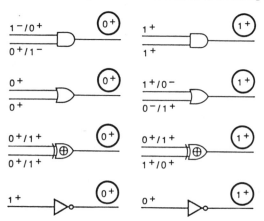

Figure 2.17 Backward sensitivity drive requirements for different gates.

partial solution and 1^+ for another partial solution. Afterwards the process is returned to the top decision block to determine whether the driving lead is the reconvergent point. When the reconvergent point is reached, the process returns to Figure 2.14 via P8 where all sensitivity value implications on all the branches are justified using the backward sensitivity and insensitivity drives rules in Figures 2.17 and 2.18. If no inconsistency is found in the justification process, the determination of the sensitivity of the stem is carried out according to the stem analysis rules in Table 2.1. The resulting enumerated values are stored as a partial solution before the process returns to Figure 2.13 via P6. After all the branches and adjacent leads are enumerated in Figure 2.13, return is made via P4 to Figure 2.12.

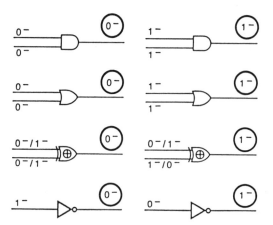

Figure 2.18 Backward insensitivity drive requirements for different gates.

Here in Figure 2.12, determination is made from the result of the block "Declare this fanout loop as redundant" in Figure 2.14 whether the Testability–Measure Analyzer should be rerun with all redundant sensitive counts set to 0. Return is then made to point P3 and the procedure is rerun without selecting those input sensitivity values for the reconvergent fanout loop, which would result in detection of the same redundancy. This may lead to the use of a multiple-path sensitization using the rules in Table 2.1. In the lowest block in Figure 2.12, TMA is rerun with all negative counts added by their offsets, necessary because one or more negative counts resulted from the left block in Figure 2.13. The counts are corrected to provide positive values. The process is then returned via P2 to the decision block in Figure 2.11 which queries whether any outer fanout loop exists within which the present reconvergent fanout loop is nested. If such is the case, the process goes to handle that outer fanout loop and the entire procedure is repeated accordingly. However, if such an outer fanout loop does not exist, then it determines whether there is another unenumerated reconvergent fanout loop which does not encompass the present fanout loop. If there is no other reconvergent fanout loop, the process ends; otherwise it repeats itself until all such fanout loops are enumerated. An example in the next section will illustrate the process flow for local enumeration just discussed.

2.4.2 An Example for Local Enumeration

In Figure 2.19 we present an example to show the essence of the local enumeration technique. The first step (i.e., step 0) is a run of TMA to determine the test count matrices for all the leads in the circuit as shown in Figure 2.19a. In this run, all leads were initialized to $N_0^+ = N_1^+ = 1$ to account for the test requirements of testing all S-A-1 and S-A-0 faults in the circuit. These results provide preliminary information (i.e., various counts) for input leads of the fanout loop as well as for leads outside the fanout loop. Local enumeration is employed to restore any missing information, with the test counts being adjusted at the end of local enumeration. Since sensitivity propagation problems exist across a fanout node, we must start a process to determine if the sensitive counts for the stem and branches are handled precisely by verifying if the sensitive values 0^+ and 1^+ can be driven along the branches to the reconvergent gate. In this process, a selected input lead for a reconvergent fanout loop is first designated as a driving lead and the sensitivity value is driven towards the reconvergent gate. Then, the individual sensitivity and insensitivity values are driven backward until all the relationships are satisfied. Note that in the process of sensitivity drive we can detect any redundancies by finding conflicts in logic value assignments, and we then removed these redundancies and rerun TMA with redundant counts set to 0 (see Figure 2.20). We will discuss the use of the local enumeration technique to detect and eliminate redundant faults in Chapter 3 together with other DFT techniques for digital circuits. After removing redundancies, all the local enumer-

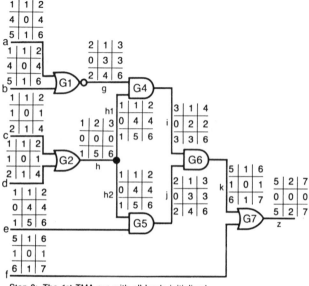

Step 0: The 1st TMA run with all leads initialized
to $n_0^+ = n_1^+ = 1$

Figure 2.19a Local enumeration example step 0: First TMA run.

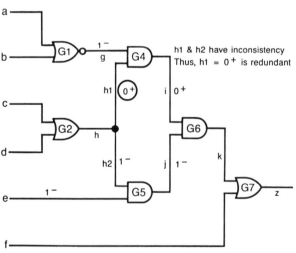

Step 1: (Beginning of local enumeration) enumerate
$h1 = 0^+$

Figure 2.19b Local enumeration example step 1: Enumerate $h1 = 0^+$.

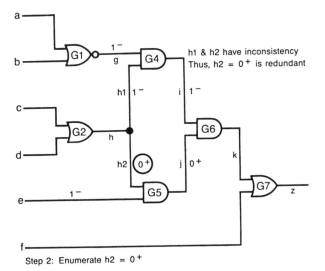

Step 2: Enumerate h2 = 0 +

Figure 2.19c Local enumeration example step 2: Enumerate $h2 = 0^+$.

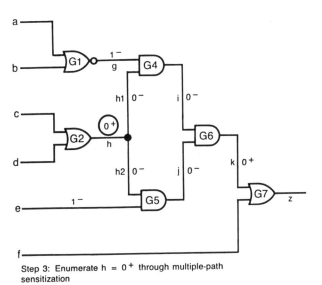

Step 3: Enumerate h = 0 + through multiple-path
sensitization

Figure 2.19d Local enumeration example step 3: Enumerate $h = 0^+$ through multiple-path
sensitization.

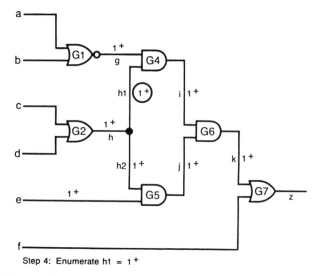

Step 4: Enumerate h1 = 1⁺

Figure 2.19e Local enumeration example step 4: Enumerate $h1 = 1^+$.

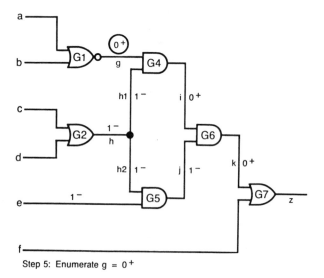

Step 5: Enumerate g = 0⁺

Figure 2.19f Local enumeration example step 5: Enumerate $g = 0^+$.

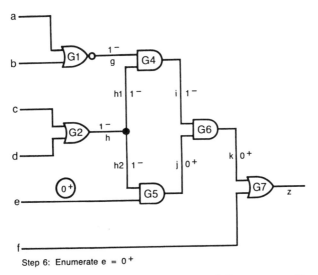

Figure 2.19g Local enumeration example step 6: Enumerate $e = 0^+$.

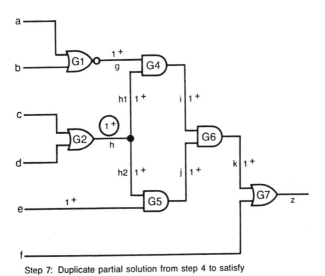

Figure 2.19h Local enumeration example step 7: Duplicate partial solution from step 4 to satisfy extra requirement on lead h.

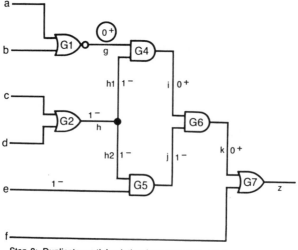

Step 8: Duplicate partial solution from step 5 to satisfy
extra req. on g (End of local enumeration)

Figure 2.19i Local enumeration example step 8: Duplicate partial solution from step 5 to satisfy
extra requirement on lead g.

ation results for the fanout loop are produced to constitute a set of partial solutions to manifest the behavior of sensitive values propagating through the loop. These sets of partial solutions will be used for outer loop enumeration or the global enumeration as part of the test generation process to be discussed in the next section.

We will step through this example here to show the essence of the local enumeration technique. Figure 2.19b is step 1 with respect to the reconvergent fanout loop between lead h and lead k. We want to enumerate both 0^+ and 1^+ on all the branches, starting with lead $h1$ having a 0^+ (encircled) by employing the forward sensitivity drive rules indicated in Figure 2.16 as well as backward sensitivity and insensitivity drive rules in Figures 2.17 and 2.18. The sensitive value 0^+ on lead $h1$ is driven forward to reconvergent gate G6 and backward resulting in the sensitivity and insensitivity values marked on the leads in Figure 2.19b associated with the fanout loop. It is seen that leads $h1$ and $h2$ require different logic values (i.e., 0 required on $h1$ but 1 on $h2$), and this assignment results in a conflict since branches from the same stem must have the same logic value. Therefore $h1 = 0^+$ is considered redundant, which means its S-A-1 fault can not be detected.

Similarly, the enumeration of $h2 = 0^+$ in step 2 in Figure 2.19c also resulted in a conflict in logic values on leads $h1$ and $h2$. Therefore $h2$ is redundant and its S-A-1 fault can not be detected. In step 3 in Figure 2.19d, the enumeration of $h = 0^+$ through multiple-path sensitization along both $h1$ and $h2$ can still be successful even though both branches are insensitive for 0 iff the reconvergent paths constitute all input leads of the reconvergent gate and they are all U-paths (see Case 1 in Table 2.1). Since all these conditions are met, the stem h can have 0^+ through multiple-path sensitization.

Next we want to enumerate $h1 = 1^+$ in step 4 in Figure 2.19e, followed by $g = 0^+$ in step 5 in Figure 2.19f and then $e = 0^+$ in step 6 in Figure 2.19g. All the results are marked on the various leads associated with the reconvergent fanout loop. In step 7 in Figure 2.19h, the partial solution from step 4 was repeated to satisfy the extra requirement on lead h of having $h_1+ = 2$ (i.e., h assumes a sensitive 1 twice) in the test count matrix for h in Figure 2.19a, since only one 1^+ has been accumulated in steps 1 through 6. Similarly, in step 8 in Figure 2.19i, the partial solution from step 5 was duplicated to satisfy the extra requirement on lead g, since the test count matrix for lead g in Figure 2.19a indicates two counts of the 0^+ sensitive value. By step 8, various sensitivity drives have been employed at the various branches and also at the other inputs of the reconvergent fanout loop until the numbers in the test count matrices are satisfied to the extent possible, taking redundancy into consideration. The solutions associated with these steps are stored.

Based on the redundancy detected in local enumeration of the fanout loop in this example, test count matrices are adjusted with the counts corresponding to the redundant values set to 0. The Testability–Measure Analyzer was rerun to adjust the remaining test count matrices accordingly based on the constraints in equations (1) through (25) in Section 2.2.

In this example the circuit has only one reconvergent fanout loop. If more than one loop is present, a local enumeration must be performed for each loop. If one loop is located within another, local enumeration is first undertaken for the innermost loop. Following local enumeration and the second run of TMA, global enumeration is undertaken to provide the test vectors required at the input leads to satisfy the test counts in the test count matrices resulting from the second run of TMA in Figure 2.20. Global enumeration is explained in the next section.

2.5   The Test Generation Process

The testability–measure method of test generation basically consists of two major subprocesses, the local enumeration and the global enumeration processes. We have illustrated the local enumeration process with a set of flowcharts to show its essence in various levels of detail and an example consistent with the steps in the flowchart. In global enumeration, essentially the same steps are followed without starting from an innermost loop. Rather the circuit is considered between the primary inputs and primary outputs, and the partial solutions are derived which describe the input vectors.

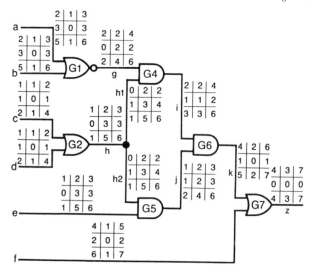

Figure 2.20 TMA Rerun results with test counts for redundant faults in example of Figure 2.19 set to 0.

In this section we will describe the global enumeration process briefly because of its similarity with the local enumeration process. The whole test generation process is then described and we will use the same example to illustrate the global enumeration as with local enumeration in the last section. Basically, we will continue with this example from where we left off in Figure 2.20 in Section 2.4.2.

2.5.1 Global Enumeration Process

The global enumeration process starts after the local enumeration is complete with the results of the rerun of TMA available. The process starts with forward sensitivity drive from selected leads using the requirements in Figure 2.16, but when a reconvergent fanout loop is encountered in the drive, the partial solutions obtained in the local enumeration are used. Referring back to Figure 2.15, the first decision block in this case is concerned with the primary output rather than a reconvergent fanout point. Once a primary output is reached, backward drive can start. In the backward drive the following rules apply:

1. If the output lead is sensitive and the gate is a reconvergent gate, a partial solution from local enumeration is selected.
2. If the output lead is sensitive but the gate is not a reconvergent gate, backward sensitive drive is perfomed according to the requirements in Figure 2.17.
3. If the output lead is insensitive and the gate is a reconvergent gate, a partial solution from local enumeration is selected.

4. If the output lead is insensitive and the gate is not a reconvergent fanout
   gate, backward insensitivity drive is performed according to the requirements in Figure 2.18.

   The process of global enumeration can be formalized as follows using a
   pseudo programming language to describe its procedural flow. Boldface
   type indicates subroutine calls.

```
Procedure global_enumeration()
{
  initialize all leads as "unset";
  for each deepest primary input lead with nonzero
  sensitive test count
    {
      if the sensitive test count is 0⁺
        then set the input lead to 0⁺;
        else set the input lead to 1⁺;
      mark that primary input lead as "set";
      decrement that test count by 1;
      current_lead = that primary input lead;
      while current_lead <> primary output lead;
        {
          if the next gate is a reconvergent fanout
            then obtain a suitable solution;
            else perform forward sensitivity drive;
          mark all leads involved as "set";
          decrement test counts of output and other
           inputs of current gate;
          current_lead = output lead of the gate;
        }
      for each gate with "set" output and "unset"
       inputs
        {
          backtrace().
          mark all leads involved as "set";
        }
    }
}
```

The procedure **backtrace** used in the procedure **global_enumeration** is described below.

```
Procedure backtrace()
  {
    if the output lead is sensitive
      then if the gate is a reconvergent gate
          pick a suitable partial solution;
      else
```

```
        {
          perform backward sensitivity drive;
           in case of ambiguity, check the test counts;
          decrement test counts on all inputs;
        }
    else if the gate is a reconvergent gate,
          then pick a suitable partial solution;
          else
           {
             perform backward insensitivity drive;
              in case of ambiguity, check the test counts;
             if the output is 0⁻ (for AND gate) or
                     1⁻ (for OR gate) do nothing; or
             decrement test counts on inputs;
           }
};
```

2.5.2 Program Description of the Test Generation Process

The complete test generation process can be formalized as a computer program as follows. With included comments and the background materials described in previous sections, the program is self-explanatory.

```
Program test_gen()
  {
     preprocessing();
     global_enumeration();
  }


Procedure preprocessing();
  {
     for every input lead
      {
         initialize the test count matrix;
         calculate its depth measured from the
          primary output;
      }
     while test count matrices are stabilized
        {
           forward_propagate();
           backward_propagate();
        }
     if there are any reconvergent fanouts
        then fanout_handling();
  }
```

```
Procedure forward_propagate();
  {
    forward propagate matrices from primary inputs
     to primary outputs;
  }


Procedure backward_propagate();
  {
    backward propagate matrices from primary outputs
     to primary inputs;
  }


Procedure fanout_handling();
  {
    for each group of related reconvergent fanout loops
        {
        /* recursively figure out the propagation
         * conditions of fanout loops, starting with
         * the innermost one. */
        from innermost loop to outermost loop do
          {
            local_enumeration();
            if test counts have changed or redundancy exists
                then reinvoke TMA to refine test count matrices;
          }
      }
  }


Procedure local_enumeration()
  {
    from deepest branch to shallowest branch (measured
     from PO)
        {
        /* verify the existence of sensitive 0 and 1
         * on the branch */
        for e_val = 0+ and 1+
          {
            if e_val has not been enumerated on the branch
              then
                {
                  verify the existence of e_val on the
                  branch with forward and backward
                  sensitivity/insensitivity drives; utilize
                  partial solutions from inner loops if
                  applicable;
                if no inconsistency
                  then
```

```
            {
                perform stem analysis to determine
                 stem logic and sensitivity values;
                record the enumeration as a partial
                 solution;
                {
            else declare the branch fault and
            equivalent faults undetectable;
              {
        }
    }
/* if branches can not be sensitized, use
 * multiple-path sensitization */
if 0⁺ or 1⁺ has not been enumerated on the stem
   then
      {
         do multiple-path sensitization;
         if it fails
          then the stem faults undetectable;
         else record the enumeration as a partial solution;
      }
     /* enumerate all adjacent leads along all reconvergent
        paths. */
      from deepest adjacent lead to shallowest adjacent lead
         {
            for e_val = 0⁺ and 1⁺
               {
                  if e_val has not been enumerated on the
                   lead
                   then
                      {
                         verify the existence of e_val on the
                         lead by forward and backward drive
                         of logic and sensitive values, utilize
                         partial solutions from inner loops if
                         applicable;
                        if no inconsistency
                         then record the enumeration as a
                          partial solution;
                        else consider the lead fault and
                         equivalent faults undetectable;
                      }
                  }
            }
      /* if local enumeration could not satisfy
       * all requirements from previous stage,
       * duplicate some of the existing partial
       * solutions instead */
```

```
            if stem and any adjacent leads have number of
            enumerations less than the corresponding sensitive
            test counts,
            then copy the existing partial solution to satisfy
            remaining test requirements;
    }
```

The procedure **global_enumeration** has been presented in the previous section and therefore is not repeated here.

2.5.3 A Complete Example to Illustrate the Test Generation Process

In this section, we present a complete example to illustrate how the test generation method produces test vectors using testability results and the local and global enumeration processes. In order to preserve continuity we use the same example that was used to illustrate the local enumeration process in the previous section. We will start from Figure 2.20 (considered as step 0), where the test count matrices for the reconvergent fanout loop are adjusted with redundancy counts set to 0. Using the Testability–Measure Analyzer, the propagation of test counts forward and backward is repeated. Thus, the test counts at leads $h1$ and $h2$ for sensitivity 0^+ have been set to 0 in each case. The test count matrices include adjustments of the remaining test count matrices during the second TMA run based on the constraints in equations (1) through (25).

Following local enumeration and the second run of TMA, global enumeration is undertaken to provide the test vectors required at the input leads to satisfy the test counts indicated in the test count matrices resulting from the second run. Referring to Figure 2.21a (step 1), the circuit is driven in accordance with sensitivity drives illustrated in Figure 2.16 from $c = 0^+$, since c and d inputs are farthest from the primary output at the lead z. Driving this sensitivity value forward and backward in accordance with the requirements in Figures 2.16 through 2.18 results in the values at the various leads shown in this figure. The partial solution from step 3 in Figure 2.19d is adopted for the reconvergent fanout loop for the lead $h = 0^+$.

In step 2 of Figure 2.21b, we drive from $c = 1^+$ and adopt the partial solution from step 4 in Figure 2.19e for the fanout region. Similarly, in step 3 of Figure 2.21c we drive from $d = 1^+$ and adopt the partial solution from step 7 in Figure 2.19h for the fanout region, while in step 4 of Figure 2.21d we drive from $a = 1^+$ adopting the partial solution from step 5 in Figure 2.19f. In step 5 of Figure 2.21e we drive from $b = 1^+$ and adopt partial solution from step 8 in Figure 2.19i for the fanout region. In step 6 of Figure 2.21f the drive is from $e = 0^+$, while the partial solution from step 6 in Figure 2.19g is adopted. In step 7 of Figure 2.21g, the drive is from $f = 1^+$, and we copy the solution from step 3 in Figure 2.21c, but with all sensitivity changed to insensitivity to satisy $k = 0^-$.

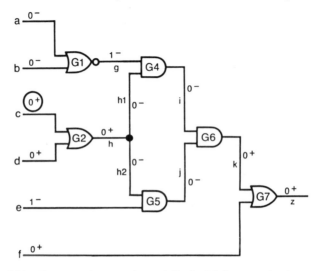

Figure 2.21a Test generation example step 1: Begin global enumeration from $c = 0^+$.

The partial solutions obtained in steps 1 through 7 in Figure 2.21 in terms of the sensitivity and insensitivity values at each lead for a separately selected input drive are stored in memory. These partial solutions satisfy the test count matrices in Figure 2.20. The various successive drives either themselves satisfy test counts or drive other leads to satisfy test counts, until the total test counts equal

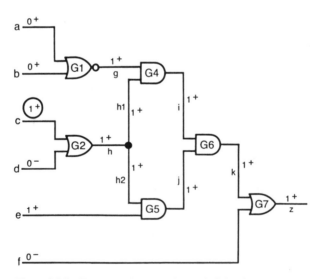

Figure 2.21b Test generation example step 2: Drive from $c = 1^+$.

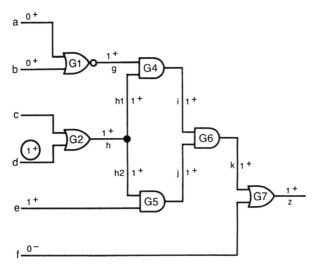

Figure 2.21c Test generation example step 3: Drive from $d = 1^+$.

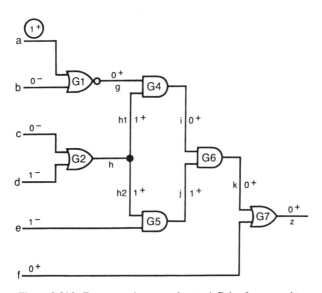

Figure 2.21d Test generation example step 4: Drive from $a = 1^+$.

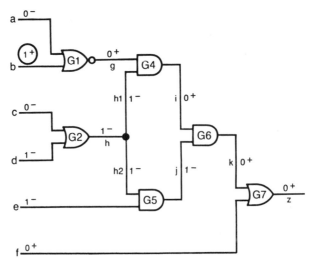

Figure 2.21e   Test generation example step 5: Drive from $b = 1^+$.

the accumulated sum of partial solutions. In every case, it is not necessary to drive every sensitivity and insensitivity value indicated by the test count matrices, because the drive by one sensitivity or insensitivity value frequently dictates the sensitivity or insensitivity values for other input leads.

The complete test-set for the circuit is illustrated in Figure 2.22. Note that the

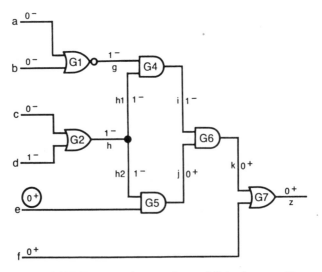

Figure 2.21f   Test generation example step 6: Drive from $e = 0^+$.

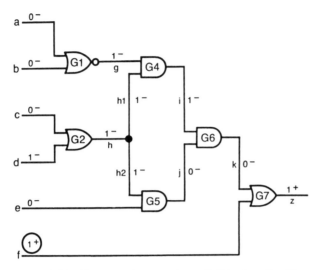

Figure 2.21g Test generation example step 7: Drive from $f = 1^+$.

seven test vectors are represented by the seven rows of values for the leads (a, b, c, d, e, and f). The sensitivity values for the different drives are marked with a circle for the convenience of comparison. Note that the total sensitivity and insensitivity values for each lead when these seven test vectors are applied should equal the test counts indicated for that lead in Figure 2.20. For instance, according to the test count matrix for the lead j in Figure 2.20, lead j assumes 0^+ once, 1^+ twice, 0^- twice, and 1^- twice. From Figure 2.22, we see that these counts (under column j) are met exactly by lead j when the seven vectors are applied. Note also that the output lead z assumes 0^+ four times and 1^+ three times, with a total of seven tests required to test this circuit. This is predicted by the Test Counting algorithm.

From this example, we see that the test generation method described uses the testability information directly to generate test vectors. The physical meaning of

A Complete Test Set

a	b	c	d	e	f	g	h	h1	h2	i	j	k	z
0⁻	0⁻	ⓞ⁺	0⁺	1⁻	0⁺	1⁻	0⁺	0⁻	0⁻	0⁻	0⁻	0⁺	0⁺
0⁺	0⁺	①⁺	0⁻	1⁺	0⁻	1⁺	1⁺	1⁺	1⁺	1⁺	1⁺	1⁺	1⁺
0⁺	0⁺	0⁻	①⁺	1⁺	0⁻	1⁺	1⁺	1⁺	1⁺	1⁺	1⁺	1⁺	1⁺
①⁺	0⁻	0⁻	1⁻	1⁻	0⁺	0⁺	1⁻	1⁻	1⁻	0⁺	1⁻	0⁺	0⁺
0⁻	①⁺	0⁻	1⁻	1⁻	0⁺	0⁺	1⁻	1⁻	1⁻	0⁺	1⁻	0⁺	0⁺
0⁻	0⁻	0⁻	1⁻	ⓞ⁺	0⁺	1⁻	1⁻	1⁻	1⁻	1⁻	0⁺	0⁺	0⁺
0⁻	0⁻	0⁻	1⁻	0⁻	①⁺	1⁻	1⁻	1⁻	1⁻	1⁻	0⁻	0⁻	1⁺

Figure 2.22 Test vectors generated by the testability–measure method.

the test count matrix on each lead is interpreted to provide the requirements on each lead in terms of the number of sensitizations and insensitizations. The correct combination of test vectors is obtained by driving sensitivity values at individual input leads forward and backward in the circuit to provide the needed sensitivity and insensitivity values for the rest of circuit. With each drive, a partial solution is formed and it constitutes a test vector. The partial solutions and the vectors are stored one at a time until all the test counts are exhausted. As a result, a set of test vectors is generated satisfying the user specified test requirements of sensitive 0s and 1s to test individual stuck-at faults. Test generation in this manner gives us a method of obtaining solutions known to exist, in contrast with most previous methods that blindly search for a solution.

2.6 Summary

In summary, a test generation method has been presented that determines a set of test vectors based on the measurement of testability of a circuit of interconnected gates according to a Test Counting algorithm. The Test Counting algorithm includes propagation of sensitivity test counts forward from inputs to an output and backward to the inputs through the intervening gates of the circuit. The test generation process can be summarized in the following steps:

1. Enumerating the test counts from the test count matrices by driving individual sensitivity values from the test count matrices of an input lead forward to an output lead and backward to the input lead in a number of successive passes in order to accumulate the test counts described by the test count matrices.
2. Separately storing a set of sensitivity values to which the input leads are driven for each pass which defines a set of test vectors for the circuit. Each pass includes driving a sensitivity value forward and backward through the circuit according to the logical determination of sensitivities appearing on the driven and adjacent leads.

The steps above first locally enumerate sensitivity values for reconvergent fanout loops to determine the sensitivity of the stems and then globally enumerate the sensitivity values for the remainder of the circuit. Test vectors thus generated satisfy the original test requirements specified by the user in the form of $N_0^+ = N_1^+ = 1$ for each fault to be detected.

The test generation method for combinational circuits described in this chapter takes an entirely different approach to solve this difficult problem. For circuits without many levels of nested reconvergent fanout loops, the test generation method introduced is quite efficient.[8] However, there is a need for more research work in the handling of multiple-leveled nested reconvergent fanout loops in a large circuit so that its computational efficiency can be improved. Furthermore, the basic approach in reconvergent fanout analysis can be used to detect redundancy in the fanout loop, and therefore may be used to improve testability of a circuit. This will be discussed in Chapter 3 with other DFT techniques.

References

1. Goldstein, L.M., and E.L. Thigpen, SCOAP: Sandia Controllability/Observability Analysis Program, *17th Design Automation Conference, 1980.*
2. Bennetts, R.G., "Design of Testable Circuits," Addison-Wesley, Reading, Massachusetts, 1984.
3. Goldstein, L.H., Controllability/Observability Analysis Program, *IEEE Transactions on Circuits and Systems* **CAS-26,** September 1979.
4. Agrawal, V.D., and M.R. Mercer, Testability measures—what do they tell us?, *Proceedings of 1982 International Test Conference,* November 1982.
5. Mercer, M. Ray, and Bill Underwood, Correlating testability with fault detection, *Proceedings of 1985 International Test Conference,* October 1985.
6. Akers, Sheldon B., and Balakrishnan Krishnamurthy, Test Counting: a tool for VLSI testing, *IEEE Design & Test of Computers,* October 1989.
7. Hung, A.C., and F.C. Wang, A method of test generation directly from testability analysis, *Proceedings of 1985 International Test Conference,* November 1985.
8. Wang, Francis, Ghulam Nuire, and Mike Brashler, An integrated design for testability system, *Proceedings of 1985 International Conference On Computer-Aided Design,* November 1985.

Chapter 3

Sequential Circuit ATVG and DFT

In this chapter, we will discuss test generation methods used for sequential circuits and design for testability techniques for making ATVG effective. As we recall from Section 1.1, a sequential circuit contains memory devices such as flip-flops, and the output of a sequential circuit is not only dependent on its present input but also on the past history of inputs. This last factor introduces another dimension into the test generation method and makes it more complex. We will begin our discussion of test generation for sequential circuits with a generalized model used to analyze the time-dependent characteristics of sequential circuits, and then we will show how this model can be adapted for automatic test generation to detect faults in a sequential circuit. We will present several efficient test generation methods in subsequent sections. We will also discuss DFT techniques that improve circuit testability to enhance an ATVG's efficiency for sequential circuits.

3.1 Introduction

3.1.1 Basic Concepts of Sequential Circuit Testing

A sequential circuit is made up of memory elements such as flip-flops and combinational logic elements such as gates. Any digital circuit containing at least one flip-flop is a sequential circuit.

The circuit shown in Figure 3.1 is a simple sequential circuit. Note that this circuit has a single input A, a single output Z, and a D flip-flop with a reset input to initialize it to a 0 state (i.e., $q = 0$). The inputs to the flip-flop are logic functions of A and Q'.

A commonly used model of synchronous sequential circuits is called the Mealy model, as shown in Figure 3.2. In this circuit the inputs are x_1 through x_n

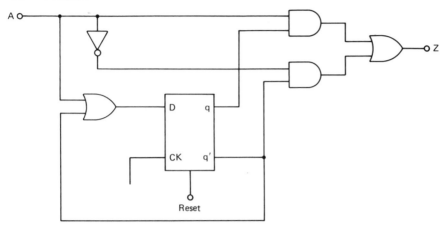

Figure 3.1 A simple sequential circuit.

and the outputs are z_1 through z_m. Note that the values of the outputs of the flip-flops represented by q_1 through q_k are called the current state S of the circuit and each value of the q output is called the current state of the corresponding flip-flop. On the other hand, the inputs to the flip-flops represented by Q_1 through Q_k are called the next state of the circuit. In synchronous mode of operation, each

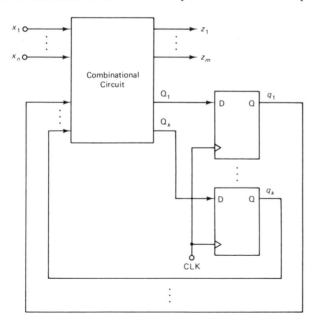

Figure 3.2 General model for a sequential circuit.

Mealy model

Table 3.1

State Transition Table for the Circuit in Figure 3.1[a]

	A = 0	A = 1
q = 0	Q = 1	Q = 1
q = 1	Q = 0	Q = 1

[a]Q = next state; A = input; q = current state.

of these flip-flops has a clock input signal that controls the state changes of the circuit. However, in an asynchronous mode of operation, the flip-flops do not have a clock input to synchronize their state transitions, and state transitions are caused by the presence of input changes. The flip-flop's output q in an asynchronous circuit is frequently called the secondary variable and its input Q is called the excitation variable.

A sequential circuit's output is determined by the combination of its input and its state. The history of previous inputs is summarized in the state S of the circuit. Note that S is a collection of values of all the flip-flop outputs at the present time, i.e., $S = (q_1, \ldots, q_k)$.

In terms of the circuit in Figure 3.1, assuming the circuit starts with the initial state of the flip-flop output $q = 0$ (i.e., $S = 0$), the next state values Q (i.e., the output of the flip-flop) corresponding to each input value of the input A for each current state $q = 0$ or $q = 1$ are indicated in the state transition table (Table 3.1). The associated output values Z for the circuit are shown in Table 3.2. The corresponding state diagram is shown in Figure 3.3 where a circle

Table 3.2

Output Table for the Circuit in Figure 3.1[a]

	A = 0	A = 1
q = 0	Z = 1	Z = 0
q = 1	Z = 0	Z = 1

[a]Z is the output.

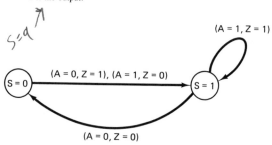

Figure 3.3 State diagram of the example circuit in Figure 3.2.

represents a state and an arc represents a transition from one state to the next state. Note that the value of the input A that caused the transition and the resulting value of the output Z are indicated on the arc from one state to the next state.

To illustrate the past-input-dependent characteristic of the output of a sequential circuit, we assume that the same circuit in Figure 3.1 has a S-A-0 fault at the output lead Z. To detect this fault, an input that causes a logic value of 1 for the fault-free circuit must be applied (see Figure 3.4). From the state diagram of the circuit in Figure 3.3, we can see that this requires $A = 0$ if the current state $S = 0$, but $A = 1$ if $S = 1$. On the contrary, if a S-A-1 exists at lead Z, then $A = 1$ must be applied if the current state $S = 0$ and $A = 0$ if $S = 1$.

Because of this characteristic of sequential circuits, a circuit must be driven into a proper state by applying a sequence of vectors in order to detect a specific fault. For instance, to detect a S-A-0 fault at the output lead of the top AND gate in the same circuit in Figure 3.4, a two-vector sequence of $V_1 = (A = 0, Z = 1)$ and $V_2 = (A = 1, Z = 1)$ is needed because the first vector is needed to drive the circuit from its power-up reset state $S = 0$ to its $S = 1$ state in order to sensitize a path from the top AND gate to the output lead Z to detect the S-A-0 fault. The second vector is used for path sensitization only. Since a synchronous circuit is involved, the input vectors must be applied in proper time relationship with the clock pulse signal of the flip-flop. It is interesting to know from the state diagram in Figure 3.3, another vector $(A = 1, Z = 0)$ can also drive the circuit from the $S = 0$ state to the $S = 1$ state. Therefore it is a viable alternative for being the first vector and it is equally effective when used with the same second vector V_2.

We can generalize what we have discussed so far to sequential circuits that are more complex ones with many inputs and outputs as well as many more states. But the basic fact to be emphasized is that the vectors must be applied in the proper sequence in order to detect specific faults in a sequential circuit as in the

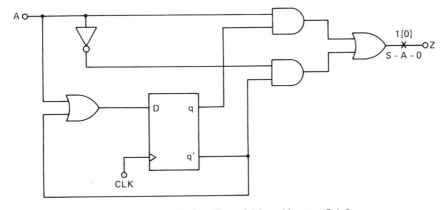

Figure 3.4 Same circuit as Figure 3.1 but with output S-A-0.

case of vectors V_1 and V_2 in the above example. We have also seen that the fault is to be detected by the last vector in the sequence as in the case of our example where the S-A-0 fault on the output lead is detected by V_2 in the sequence when the expected output at lead Z is a 1 for the fault-free circuit and the actual output is 0 (as indicated in the square bracket) for the faulty circuit.

3.1.2 A Model for Sequential Circuit Test Generation

As we have seen in the previous section, to test a specific fault for a sequential circuit, a sequence of vectors is frequently needed because of the fact that the circuit's output is dependent on both the present input and the past history of inputs applied to the circuit under test (CUT). These vectors in a test sequence perform two functions: drive the circuit under test into a proper state and detect the given fault from that state. In other words those vectors performing the first function set up the proper past inputs in order for the fault-detecting vectors in the sequence to sensitize a path to an output to detect the fault.

To represent the different states associated with a sequential circuit, the time scale is divided into time frames. Each time frame represents the circuit at a different state. A sequence of test vectors is thus generated, for each state that drives the CUT from one state to the next until the required state is entered. When generating test vectors for a sequential circuit, a circuit is frequently modeled as an iterative array of instances of the same circuit to represent this multiple-time-frame approach. Each member of the array represents the CUT behavior at a particular time frame (state) as shown in Figure 3.5, where a sequential circuit is represented by a combinational circuit (CC) and a set of flip-flops controlled by a clock. This model and the one in Figure 3.2 are equivalent in the sense that the input and output relationship for the same circuit is preserved and also in that we used the same symbols in the two models.

In a given time frame (n) of this model, a flip-flop stores the output value of

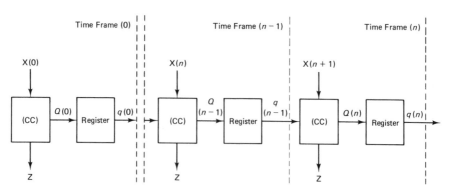

Figure 3.5 Model for sequential circuit ATVG.

the CC. Its own output $q(n)$ is the value of the present state (n). Note that within a given time frame, the circuit behaves like a combinational circuit, but the state of the flip-flop depends on both the input of the current time frame and the state of the previous time frame. Therefore, there exists a special relationship between the state values across two adjacent time frames; here the current state value $q(n)$ is determined by the combination of the current input $X(n)$ at the time frame (n) and the state value $q(n - 1)$ at the previous state. Since we are assuming the flip-flops are D type, Table 1.4 in Section 1.1.3 can be used to determine the next state value.

Each input X in a time frame is a vector and if a sequence of vectors $X(0)$, $X(1)$, . . . , $X(n - 1)$, $X(n)$ detects a particular fault, then it is a test vector sequence for that fault. When applying these test vectors, their proper ordering in the sequence must be preserved because each vector drives the CUT from a particular state to the next. In this case $X(0)$ should be applied to the CUT first, followed by $X(1)$, then $X(2)$, . . . , $X(n - 1)$, and then the last vector $X(n)$.

It should be noted that these vectors form a sequence for a given fault(s) and this order in the sequence has to be preserved in actual testing. To completely test a circuit, many such sequences have to be generated and the ATVG has to use the last state in the previous sequence as the starting state for this sequence. At the beginning stage of test generation, the ATVG needs to know the power-up state of the CUT so that it can search for a sequence of vectors, one for each time frame, starting from the power-up state to the next state, then to the last state in sequence to detect specific faults. If the information about the power-up state is not available, then the ATVG has to search for an initialization sequence of vectors to drive the circuit from any state known or unknown to a known state. In Figure 3.6 we show a typical test vector set for a sequential circuit that contains many test sequences, each for a specific fault(s).

In summary, this model allows us to search for a vector of a given time frame using the same techniques as those used in generating vectors for combinational

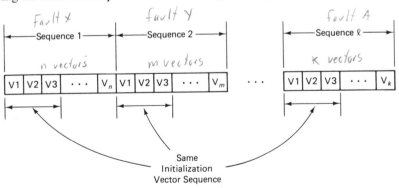

Figure 3.6 A typical test vector set.

circuits. But the logic value assignments associated with state values in a given time frame must be carried across time frames as the required logic values to be justified in the adjacent time frame. This approach simplifies the ATVG for sequential circuits and makes the process more like that of a combinational circuit. However, one more dimension (i.e., time dimension) is added to the already difficult problem. There are also other unique problems associated with sequential circuit test generation and we will discuss them together with design-for-testability techniques in later sections of this chapter.

3.1.3 A Brief History of Sequential Test Generation

The D-algorithm discussed in Section 1.5 is the pioneering technique for generating test vectors for combinational circuits. The original approach was later extended to generate tests for sequential circuits[1] using the test generation model discussed in the previous section. However, the extended approach does not enjoy the same status for sequential circuits as for combinational circuits, due to the large computer memory required in its implementation[2] and the problem of extensive backtracking. The backtracking problems sometimes occur late in the justification process[3] and any recovery process may have to span across multiple time frames. This can seriously hurt the efficiency of the AVTG using this approach.

A few more efficient algorithms for generating test vectors for sequential circuits have since been available and become more popular. Some of these techniques use reverse timing[4,5] processing to avoid the need to add extra vectors after the test generation process is complete, while others use better fault modeling techniques to reduce backtracking.[6] Recent implementations of reverse time processing for ATVG also use heuristics[7] to minimize the steps needed to recover from conflicts during backtracking. There are also other quite different approaches such as using a fault simulation[8] program to generate test vectors. In the next three sections, we will discuss three of these algorithms: the Extended Backtrace algorithm, the BACK algorithm, and the CONTEST method.

3.2   The Extended Backtrace Algorithm

3.2.1 Basic Concept

Test generation for a fault in a sequential circuit involves two steps: fault effect propagation and state initialization. Fault effect propagation is to determine a state into which the circuit must be driven so that additional vectors can be applied to make the fault observable at a primary output. On the other hand, state initialization is to determine a sequence of vectors that can drive the circuit from any state to the state specified for fault effect propagation. Both forward or reverse time processing can be used to generate tests for sequential circuits.

Forward processing generates vectors in the same order as they are applied, while reverse processing generates vectors in the order opposite to their application.

The Extended Backtrace (EBT) algorithm was developed by R. Marlett[4,5] and it generates a sequence of test vectors for a fault in a reverse time fashion (i.e., the first vector generated in the sequence is the last one to be applied in actual testing and the last vector generated is the first one to be applied). In the test generation process for a selected fault, a pair of vectors for two adjacent time frames is always considered together. The first vector in each pair considered in test generation is called the current vector (CV), while the second one considered is called the previous vector (PV). The CV is generated to propagate the fault effect to an observable primary output (PO) of the circuit so that the fault can be detected. The purpose of having a PV associated with a given CV is to set up the logic values in the circuit to satisfy the conditions imposed by the CV because the CV is associated with the current time frame while the PV is associated with the previous time frame. Therefore, in actual testing the PV is applied before its associated CV in time. The key concept for EBT is that in the process of generating the CV it automatically places requirements (i.e., logic value and state value assignments) on the associated PV because the next state value (i.e., required by the CV) is dependent on the current state (i.e., requirements imposed on the PV) and the input for a sequential circuit. As a result of this imposed relationship between the two vectors, the PV is already partially specified based on the needed state and logic values for the CV. Repeating this process in an iterative manner, a sequence of vectors can be generated by making each PV a new CV, and a corresponding new PV is added to form a new pair of vectors. This pairing can be repeated as many times as needed in order to generate a sequence of vectors that will not only drive the circuit from one state to the next until reaching the state required by the first CV but also detect the given fault. Note that the vectors must be applied in the reverse order from the generation process.

In terms of the model discussed for test generation in the previous section for a sequential circuit, the CV can be considered as the vector $X(n)$ associated with time frame (n) and PV the vector $X(n - 1)$ associated with time frame $(n - 1)$. The EBT algorithm requires in each time frame to backtrace from the output of the circuit to its primary inputs, meanwhile determining and assigning input logic values to satisfy the required output value for each device, one at a time. These input values, in turn, become the required output values of the preceding devices so the process repeats until a set of values for the primary inputs are found. In the case that the input value for a logic device comes from the output of a memory device such as a flip-flop, requirements are placed on the flip-flop's input value and the previous state value. To complete test generation this process is repeated until no more previous state value needs to be specified.

In summary, EBT involves reverse processing both in space and in time by determining the required primary input signal values for a given time frame and then carrying the state requirement backward in time until a sequence of vectors are generated to detect the selected fault. In the next section, a flowchart is presented to illustrate the steps to be followed in generating tests.

3.2.2 Flowchart Description of EBT

Figure 3.7 shows the main procedural flow of the Extended Backtrace algorithm to generate a sequence of test vectors for a given fault. For complete test generation for all faults, the procedure in the flowchart is repeated as many times as necessary. We will explain each block in the flowchart below.

The procedure for test generation starts from block 1 that indicates EBT must select a fault which has not been tested yet, then associates it with the output signal of a logic device. If this output signal is not a primary output, then a sensitized path called the topological path (TP) is selected in block 2 to allow the fault effect to propagate from the fault site to a primary output. Note that the TP in general consists of many signals that lay in the sensitized path. In block 3, two vectors are created and the components of each vector are the signals in the circuit under test. All the components of the vector are first initialized to Xs. The CV is associated with the current time frame and the PV is associated with the previous time frame.

The ultimate objective here is to build up the CV and PV in an iterative manner as indicated in block 4, which is further expanded into the flowchart in Figure 3.8. Once "Build vectors CV and PV" is completed, the CV is saved in the vector sequence and the PV becomes the next current vector (block 4). A new PV is then initialized to all Xs (block 5) and the flow goes back to point P1 so that the next pair of CV and PV are built as required.

Figure 3.8 shows the key steps in the EBT algorithm in generating test vectors. Block 1 indicates that it processes the logical devices in the TP in the reverse order from the circuit primary output back to the fault site. It assigns input values of a device in such a manner that the output of the device is sensitized. The assigned input value is then added to the CV and PV (block 2). Since this process is iterative, a decision block is then entered to determine if all signal values in the CV have been assigned. If all signals have been assigned, then the process goes to the block to save the CV into the vector sequence. Otherwise, a signal already in the CV is selected for processing.

If the signal selected is a primary input, then the same decision block as mentioned above is entered to determine if all signals in the CV have been specified. Otherwise, a decision is made whether the signal is part of the TP. If so, an input value of the device that is connected to the selected signal is assigned to sensitize the selected signal (block 4). In another words, an input signal value already assigned for a device is used to allow assignment of the next input signal (in its preceding device) to further sensitize the path from the PO to the fault site so

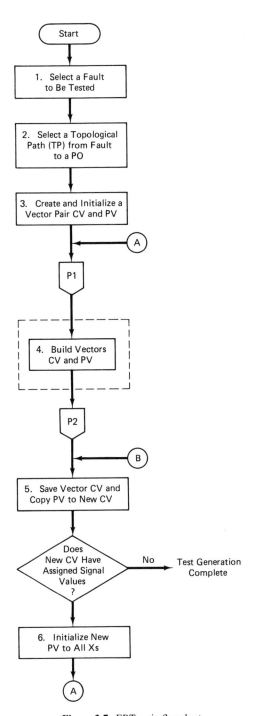

Figure 3.7 EBT main flowchart.

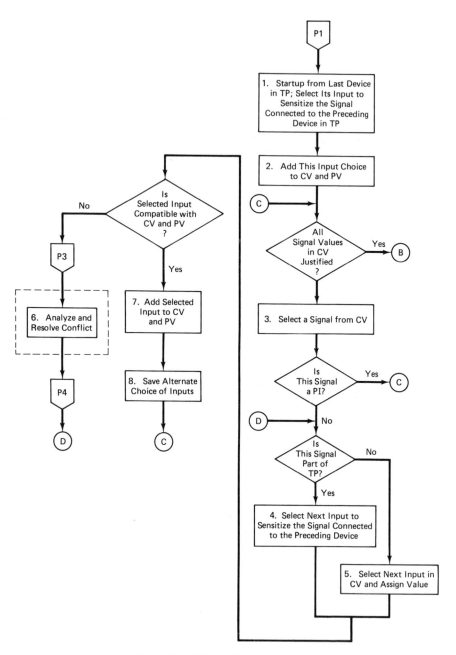

Figure 3.8 EBT "build vector" flowchart.

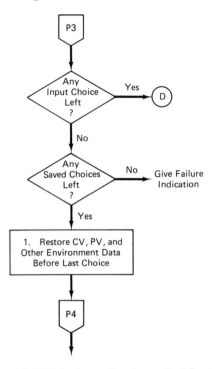

Figure 3.9   EBT "analyze and resolve conflict" flowchart.

that a continued chain of sensitized signals can be formed for fault effect propagation. A check to determine if assigned values cause conflict with signals already existing in the CV and PV is performed at this time. If there exists no conflict, these are added to the CV and PV(block 7); otherwise the process goes to the block "Analyze and resolve conflict." Note that in block 8, alternate choices for the last input assignments should be saved for potential usage in backtracking.

Figure 3.9 expands the block "Analyze and resolve conflict" from Figure 3.8. Note the first decision block in Figure 3.9 is to test if there are any other input choices left for the case that the selected signal in block 3 of Figure 3.8 is not in the TP. On the other hand, the second decision block is to determine if any other choices saved in block 8 in Figure 3.8 are left in order to backtrack to the last choice, which will require restoration of previous signal values already added to the CV and PV.

3.2.3 An EBT Example

We will use the same example as in Figure 3.4 to illustrate the concept and procedure used by EBT to generate test vectors. Note in Figure 3.10 that we

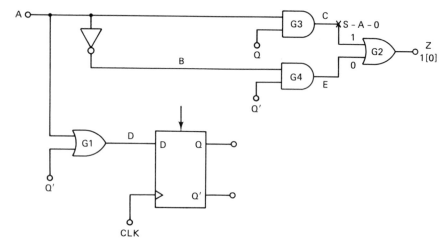

Figure 3.10 An example for EBT.

have redrawn the circuit and have also taken out the reset input to the D register in this circuit because the vectors generated by EBT can inherently initialize a sequential circuit.

We will list the steps below to show how a sequence of vectors are generated to detect a specific fault.

1. Select the stuck-at-0 fault of signal C for test generation.
2. Select a topological path (TP) from C to signal Z through the last OR gate.
3. Create current vector CV0 and previous vector PV0 and initialize them to all Xs.

	CLK	A	B	C	D	E	Q	Q'	Z
CV0	X	X	X	X	X	X	X	X	X
PV0	X	X	X	X	X	X	X	X	X

4. Build vectors CV0 and PV0.

   a. Assign a 1 to the top input of the OR gate $G2$ to sensitize C to its output.

   b. Add this input choice to CV0 and also add the corresponding output signal value for Z.

	CLK	A	B	C	D	E	Q	Q'	Z
CV0	X	X	X	1	X	X	X	X	1
PV0	X	X	X	X	X	X	X	X	X

c. Select signal value $C = 1$ from CV0.

d. Assign input signal $A = 1$ for $G3$ to sensitize A to C.

e. Add $A = 1$ to CV0.

	CLK	A	B	C	D	E	Q	Q'	Z
CV0	X	1	0	1	1	X	X	X	1
PV0	X	X	X	X	X	X	X	X	X

f. Select signal value $Z = 1$ from CV0.

g. Assign input signal $E = 0$ as required to sensitize the TP and add it to CV0.

	CLK	A	B	C	D	E	Q	Q'	Z
CV0	X	1	0	1	1	0	X	X	1
PV0	X	X	X	X	X	X	X	X	X

h. Select signal value $C = 1$ from CV0.

i. Assign input signal $Q = 1$ as required to sensitize $C = 1$ in the TP.

j. Add $Q = 1$ to CV0 and $D = 1$ (required as input in the previous time frame in order to produce $Q = 1$ in the current time frame) to PV0.

	CLK	A	B	C	D	E	Q	Q'	Z
CV0	X	1	0	1	1	0	1	0	1
PV0	X	X	X	X	1	X	X	X	X

k. Assign CLK $= 1$ to CV0 and CLK $= 0$ to PV0.

	CLK	A	B	C	D	E	Q	Q'	Z
CV0	1	1	0	1	1	0	1	0	1
PV0	0	X	X	X	1	X	X	X	X

5. Save vector CV0 and copy PV0 to new current vector CV1.

	CLK	A	B	C	D	E	Q	Q'	Z
CV1	0	X	X	X	1	X	X	X	X

6. Initialize a new previous vector PV1 to all Xs.

	CLK	A	B	C	D	E	Q	Q'	Z
PV1	X	X	X	X	X	X	X	X	X

7. Build vectors CV1 and PV1.

   a. Select signal value $D = 1$.

   b. Select and assign $A = 1$ to satisfy $D = 1$.

   c. Add this input to CV1.

	CLK	A	B	C	D	E	Q	Q'	Z
CV1	0	1	0	X	1	X	X	X	X
PV1	X	X	X	X	X	X	X	X	X

8. Since all signal values in CV1 have been justified, save vector CV1 and copy PV1 to CV2.

	CLK	A	B	C	D	E	Q	Q'	Z
CV2	X	X	X	X	X	X	X	X	X

9. Since the new CV2 does not have any assigned signal values and therefore it does not require any signal conditions in the PV2 for the previous time frame, we conclude that no more test vectors are needed to detect the signal C S-A-0 fault. The test vectors CV0, CV1, and CV2 should be applied in the reverse order in testing from the order they were generated, i.e., V0 (CV2) first, then V1 (CV1), and V2 (CV0) last. The three test vectors V0, V1, and V2 are listed below.

	CLK	A	B	C	D	E	Q	Q'	Z
V0	X	X	X	X	X	X	X	X	X
V1	0	1	0	X	1	X	X	X	X
V2	1	1	0	1	1	0	1	0	1

10. This completes the test generation process for this particular fault.

Note that in the above example, the vector V0 is a null vector since it does not contain any required signal values and it is used as an indicator to show the beginning of a sequence of vectors for a particular fault.

3.3   The BACK Algorithm

We have just used a very simple example to show the essence of the EBT algorithm to generate test vectors for sequential circuits. In a complex sequential circuit, EBT may have to try many paths in order to find a sensitized path TP to propagate the fault effect to a primary output. This can become impractical if EBT has to try path by path for a very large number of possible paths. We will discuss some other significant ATVG methods which have alleviated this problem.

3.3.1 The Basic Concept

The major differences between EBT and the BACK algorithm[6] are the methods used in propagating the fault effect and the circuit models used.[2] The BACK algorithm preselects a sensitized primary output instead of a sensitized path for fault propagation. It also uses the backward justification process as in the EBT algorithm. A sensitized path is created automatically in the BACK algorithm when the justification of the sensitized primary output is complete. To make the proper selection of a sensitized primary output, the BACK algorithm uses a testability measure called drivability to predict which primary output is to have a sensitized value. Since the number of primary outputs is usually limited in a circuit, this algorithm can try all the primary outputs and eventually detect all testable faults.

The drivability measure is an indication of the degree of difficulty to propagate the fault effect from the fault site to a particular signal in the circuit. Two values are used, one for 0/1 sensitization and one for 1/0 sensitization. The drivability values calculated for all the primary outputs are then used as the guidance to select the most promising primary output for fault effect propagation. The drivability values are based on SCOAP-like controllability and observability measures.[9] Some other testability measures can also be used for the same purpose.

Three steps are taken to calculate the drivability for a fault.

1. Calculate the controllability value for each signal in the fault-free circuit and also each signal in the faulty circuit.
2. Using the controllability value at the fault site which causes a value opposite to the stuck-at-fault value as the drivability value at the fault site.
3. Propagate the drivability value at the fault site forward to all the primary outputs.

The PO associated with the smallest drivability value should be selected first. The remaining values should be selected starting with the next smallest drivability value if the first choice fails to sensitize the fault to this PO. This process continues until either a test is found or all primary outputs have been exhaustively tried. In the latter case, there does not exist a test for this fault.

3.3.2 Flowchart for the BACK Algorithm

The procedure of the BACK algorithm is summarized in the flowchart in Figure 3.11.

3.3.3 The SPLIT Model

The BACK algorithm also uses a SPLIT circuit model[2] which helps it to speed the line justification process. The basic idea is to split the logic values (i.e., 0,

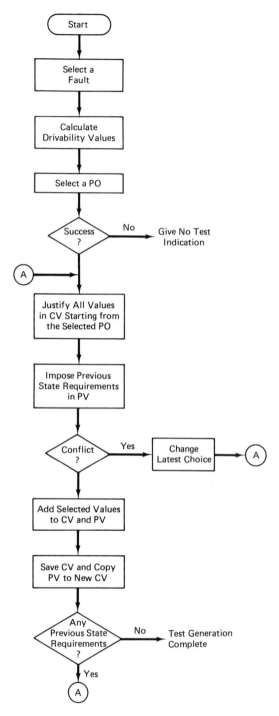

Figure 3.11 The BACK algorithm flowchart.

1, and X) for a signal into two sets of values, one for the fault-free circuit and one for the faulty circuit, so that they can be treated separately. A relation exists between each of these two sets of values. The relation can be unknown, difference, and equivalence. Difference means that the values must be different in the faulty and the fault-free circuits. Equivalence means that the value must be the same in the two circuits. Unknown means the values have an unknown relationship. With the values of a signal for the faulty and for the fault-free circuit treated separately and using the relation to guide the process, justification can be done much faster.

From published results, the BACK algorithm implemented with the SPLIT model in an ATVG outperformed the EBT algorithm in a benchmark study.[10,6] It not only requires lesser run-time computer memory, but it is also more efficient in terms of shorter time to run the ATVG and higher fault coverage.

3.4 A Simulation-Based Method: CONTEST

As we discussed in Figure 1.17 of Section 1.4, the test generation process normally uses a fault simulator to verify the test vectors generated. The ATVG either uses a stand-alone or its own fault simulator to verify the faults detected. It can also detect any timing problems caused by the zero-delay assumption in ATVG for the combinational logic part of the circuit. More recently, approaches that use simulation to generate tests have been proposed,[11] and they have been proven successful in generating tests for both synchronous and asynchronous sequential circuits.

3.4.1 The Basic Concept

The basic idea behind the simulation-based test generation method is to simulate a selected initial vector and calculate a cost function based on its simulation results. For instance, a typical cost function is the total number of unknowns (i.e., Xs) at the outputs of a sequential circuit where all outputs are assigned as unknowns at the beginning of the simulation process. A trial next vector is then selected by modifying the initial vector, and the cost function is calculated from the simulation results for this trial vector. This cost is compared with the initial cost. If the cost is decreasing, the trial vector becomes the next vector. Otherwise the initial vector is progressively modified until a decreasing cost function is achieved. This iterative process is done successively by progressive vector modification, resimulation, and repeated cost function calculation until a sequence of test vectors is found in which the cost eventually drops below a specified threshold. It also takes consideration of circuit delays in simulation like a conventional event-driven simulator and therefore the vectors generated do not cause race and hazard problems when applied in an ATE to actually test the circuit.

In this section, we will present the CONTEST[8] method of test generation, which is effective for both combinational and sequential circuits. CONTEST can generate test vectors for a group of faults, for a single fault, and also for the initialization sequence of a circuit. However, separate cost functions are used for these three applications.

3.4.2 Concurrent Test Generation for a Group of Faults

The CONTEST method considers a group of faults simultaneously at the early stage of test generation when a large number of faults have not been tested. This can generate tests for multiple faults at one time to quickly reduce the number of faults to be considered.

The cost function for each fault in a group is taken as the number of gates on a path between the fault effect and a primary output. In case a flip-flop exists on the path, a higher cost value is assigned than that for a gate because a flip-flop requires both setting its data input and also clocking it in order to propagate the fault effect. When the fault effect is propagated to a primary output, the cost drops to zero and the fault is considered detected. The objective in test generation thus becomes that of finding successive vectors to propagate the fault effect from one logic device to the next until a primary output is reached.

In evaluating a trial vector, costs associated with all the faults in a group must be considered jointly. A simple rule used to determine the acceptance of the trial vector is to see if the combined cost of a percentage of the lowest cost faults is decreasing. Once the test generation process has determined that the cost functions can not be reduced any more, it automatically switches to another type of cost function associated with a single fault test generation.

The process starts with selecting an initial vector of all 0s or any user specified vector(s). A trial vector is obtained by modifying one input bit at a time in the last vector of the sequence until its calculated cost is lower than that of the last vector. It is then added to the sequence of vectors and itself becomes the new last vector. This process continues until all costs drop to zero or no new reduction of costs is possible. This phase of test generation terminates and the single fault test generation takes over if there remain undetected faults.

Note that in generating tests for a synchronous circuit, the clock signal must be specified. The primary input signals are applied in simulation at the transition of the clock signal. For an asynchronous circuit, simulation starts after the primary input signals have changed.

3.4.3 Single-Fault Test Generation

To generate tests for a single stuck-at fault, a sequence of vectors must first be found to set the signal value opposite to the faulty value at the site of the fault. This is called the fault activation sequence. Another sequence of vectors must then be found to sensitize a path to propagate the fault effect from the fault site to a primary output. This is called the propagation sequence.

Two cost functions are defined, one for fault activation and one for fault effect propagation. These functions are based on SCOAP-like controllability and observability measures.[9] The activation cost is defined as the controllability value for setting the signal value at the fault site but opposite to the stuck-at value of that signal. Note that once a circuit enters into a state where the fault is already activated, this cost drops to zero. The propagation cost is defined as the observability value for the signal at the fault site. The cost of a single-fault test generation is a weighted sum of these two costs. The test generation is identical to that of a group of faults except the cost functions that are used to determine if the trial vector should be accepted are different.

Results of test generation indicate that CONTEST performs better than EBT and other methods. However, CONTEST sometimes generated more vectors to test a circuit than other methods, because of the single-bit-modification heuristic used in selecting adjacent vectors. Another significant application of the simulation-based test generation approach is to find a sequence of vectors to initialize a sequential circuit from an unknown state to a common known final state. This is one of the topics we will discuss in the next section.

3.5 DFT for Sequential Circuits

We have reviewed in the previous sections several effective algorithms and methods for automatic generation of tests for sequential circuits. However, there is no single ATVG product today that can handle any arbitrarily complex circuits, especially if there exist inherent testability problems in the design of a circuit. These problems can render the most effective ATVG useless. We summarize these critical testability problems frequently encountered in digital circuits as follows.

1. A noninitializable design
2. Effects of component delays
3. Nondetectable logic redundant faults
4. Existence of illegal states
5. Oscillating circuits

These problems are not the only ones that can make a circuit untestable, but they certainly are the major causes for an ATVG to fail to generate tests. We will identify DFT approaches to alleviate them at the early stage of circuit design.

3.5.1 Design Initializable Circuits

Since a sequential circuit contains memory devices such as flip-flops, the circuit is frequently in an unknown state when the power is first turned on (i.e., power-up). A sequence of test vectors is needed to drive the circuit from any unknown states into a state required for propagating the fault effect to a primary output of the circuit. Depending on the design, such a sequence of vectors may exist for a circuit. If so, we then call it initializable. But sometimes the design precludes the existence of such a sequence; we then call such a circuit noninitializable.

The circuit in Figure 3.12a is initializable because we can apply a sequence of

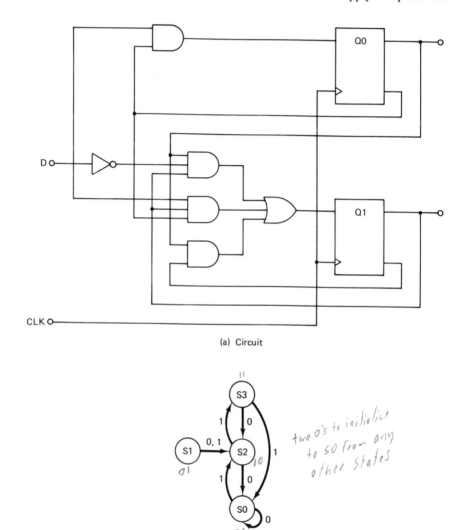

(a) Circuit

(b) Its FSM

Figure 3.12 An initializable circuit.

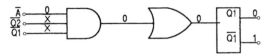

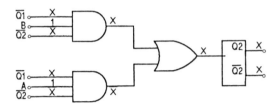

Figure 3.13   An noninitializable design applying vector V_1 ($A = 1$ and $B = 1$).

two 0s to the input to drive it from any state to the S0 state based on its FSM representation in Figure 3.12b. On the other hand, the circuit in Figure 3.13 is noninitializable. If we first apply the vector ($A = 1$ and $B = 1$), the resulting outputs are $Q_1 = 0$ and $Q_2 = X$, where X represents an unknown value. If we apply the second vector ($A = 0$ and $B = 0$), then the output becomes $Q_1 = X$ and $Q_2 = 0$ as shown in Figure 3.14. The reader can try other sequences of

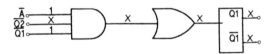

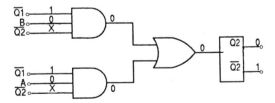

Figure 3.14   The same noninitializable design as in Figure 3.13 applying vector V_2 ($A = 0$ and $B = 0$).

vectors to see if a sequence of vectors can be found that will bring the circuit into a known state (i.e., neither Q_1 nor Q_2 having an unknown value). As you will find out, this design is noninitializable because no sequence of vectors can be found to bring it from any unknown state to a known final state.

At this stage, we may want to ask "why do we need to design initializable circuits?" This is because initializable designs can facilitate circuit testing as well as design verification. In testing there are three major advantages to having initializable circuits.

1. It is possible to generate vectors without the need to know the power-up state because the initialization sequence can drive the circuit into a known state no matter what its power-up state. For a noninitializable circuit, it may need to take an actual measurement of the state of the circuit at power-up before automatic generation of test vectors.
2. A large test set for a circuit containing many sequences of test vectors is frequently generated for a complex circuit and it may exceed the vector length (called burst length) of some ATE. In this case the original test-set has to be divided at the sequence boundaries into manageable subsets of tests to be applied one subset of vectors at a time on the ATE. In this case the initialization sequence is needed at the beginning of each subset of vectors (as shown in Figure 3.6) to initialize the circuit from the last state in the previous sequence to a common known state.
3. For in-circuit testing of a circuit board, it is sometimes impractical to turn the power off and then back on as required by a noninitializable circuit to match its power-up state with that which was used during automatic test generation.

The easiest way to make a circuit initializable is to change the design so that it uses a flip-flop with a master reset or clear input signal which can set it to a 0 state asynchronously. The noninitializable circuit in Figure 3.14 can be modified by using two D flip-flops each with a reset input. At the power-up time, the reset input goes high and the circuit always makes the transition from any state to the 00 state. This modification requires a dedicated external pin on the circuit for initialization but it is worth it from a testability point of view.

Next we will present a simple and easy method to generate an initialization sequence of vectors for a circuit. This method uses a logic simulator to find a sequence of inputs that reduces a predefined cost (e.g., the total number of unknowns at the primary outputs of a sequential circuit) down to a minimum. As shown in the flowchart in Figure 3.15, the test generation process starts with assigning all flip-flops in the circuit to the unknown state and an initial cost C_0 defined to equal the number of flip-flops to be initialized. Then an all 0 vector is selected as the current vector for simulation. From simulation results we can determine the number of flip-flops still remaining in the unknown state (i.e.,

Handwritten annotations:

Cost $C_0 = n$ # of unknowns, # of primary outputs

$C_1 = (n-m)$ $m = 0, 1, \ldots n$

change vectors, by one bit

c7 until # unknowns = 0

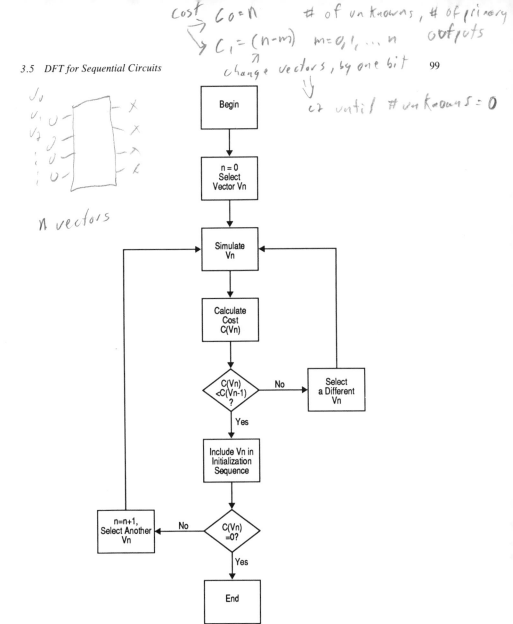

Handwritten annotations (left margin):

J_0
V_1 U —
V_2 0 —
$\vdots$ 0 —
0 —

n vectors

Figure 3.15 A simulation-based method to generate an initialization sequence.

having X values) and let this be the current cost. If the current cost is lower than the ori-ginal cost, then save this cost as the new original cost and this vector as part of the initialization sequence. Otherwise, the current vector is modified by changing one of its bits and resimulated until a vector with a lower cost function is obtained. This process is repeated by modifying each successive vector one bit at a time from its preceding vector until a lower cost function is obtained. The whole generation process stops when the cost drops down to zero (i.e, all flip-flops are initialized) or no more input can be found that reduces the cost any further.

There are other methods [12] that require exhaustive search of possible states of a circuit in order to find an initialization sequence. However, we found that the simulation-based method is very easy to implement if a logic simulator is available and that it is also very effective [13] based on our experimental results with sequential PLD circuits. If the design is initializable, it usually took much less computer time to find an initialization sequence than other methods. For noninitializable designs, it also terminated its process much sooner because it searches only n out of 2^n possible neighbors of a given vector (n is the number of primary inputs).

We encourage a design engineer to use this method to verify if his or her design is easily initializable at the early stage of design so that initializability can still be improved. Once an initialization sequence is found, it should also be transferred as part of the design verification vectors to the test engineer. In case a given circuit has already reached the manufacturing stage without any initialization vectors, the test engineer can also independently generate the needed vectors to initialize the circuit using the method just discussed.

3.5.2 Design to Minimize the Effects of Timing Delays

In sequential circuits, timing delays associated with hazard and races in a circuit can cause problems for ATVG. Sometimes an ATVG fails to generate vectors under these conditions and at other times it generates bad vectors (i.e., these vectors fail for a fault-free circuit on an ATE). These problems are frequently caused by an ATVG's ignorance of unequal delays along two or more signal paths in searching vectors because it normally uses a zero-delay model (i.e., assuming all the gates are ideal and do not introduce any delays). In practice, gates and wires in a circuit introduce propagation and transmission delays. If of sufficient values these delays can cause not only test generation problems but also circuit malfunction, and they must be handled properly at the design stage.

In asynchronous circuits, delays in the feedback paths are seldom equal. Thus, when two or more secondary variables (see definition in Section 3.1.1) must change values simultaneously to bring about a state transition from one stable state to another, these secondary variables go through some intermediate values before settling down into their final values. Due to unequal delays a critical race may exist, i.e., some of these variables may change faster than others and the net result is that the circuit may end up in an incorrect state. Failing to recognize the existence of the critical race condition, the vectors generated assuming zero delays will normally predict incorrect states in both the initialization and fault

Race & Hazard

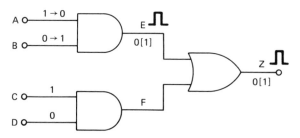

Figure 3.16   Hazard condition in a circuit.

effect propagation sequence, and these vectors will fail in testing a fault-free circuit on an ATE.

There exist several techniques [14] to assign state variables so that all state transitions allow only one state variable change at a time. Design engineers should take elaborate time in their design to assign proper state vectors to adjacent states.

A second problem due to unequal time delays is called hazard. A hazard exists if in a circuit, as a result of input changes, a spurious signal value appears at a gate's output, which should not occur. A synchronous circuit is normally not impacted by a hazard because the transitions are controlled by the clock. However, an asynchronous circuit may make an incorrect transition into a wrong state because of the momentary signals produced by the hazard. A simple example is given in Figure 3.16. There exists a hazard in this circuit when signals A and B are changing in the opposite direction in two adjacent vectors (while signals C and D remain constant), but signal B is faster than signal A and therefore causes a glitch in signal E and signal Z, where square brackets are used to indicate the erroneous value of the signal due to these hazard problems. If Z is connected to a clocked flip-flop, then the glitch can be filtered out by the clock signal without causing any harm to the rest of the circuit.

If this circuit is modified to now include a feedback loop from the output Z to the input of the lower AND gate, it becomes an asynchronous circuit as shown in Figure 3.17. If we apply the same two vectors used for the circuit in Figure 3.16 to the inputs of this circuit (assuming Z has an initial value 0), then the glitch caused by the unequal delays in signals A and B is latched at the output Z as shown in Figure 3.17b. This erroneous signal will make the vectors generated by ATVG fail on an ATE. Actually this type of circuit is called a combinational memory circuit because it can store a signal at its output just like a memory device.

There are two ways to circumvent this problem. The first one is to allow an ATVG to treat signal C as the clock signal and then apply the clock signal to the circuit in separate vectors only after the glitch has disappeared. This requires the design engineer to clearly record in the design all the equivalent clock signals.

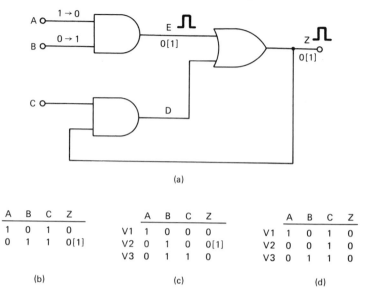

	A	B	C	Z
V1	1	0	1	0
V2	0	1	1	0[1]

(b)

	A	B	C	Z
V1	1	0	0	0
V2	0	1	0	0[1]
V3	0	1	1	0

(c)

	A	B	C	Z
V1	1	0	1	0
V2	0	0	1	0
V3	0	1	1	0

(d)

Figure 3.17 (a) An asynchronous circuit; (b) vectors that latch the incorrect output; (c) resolve hazard by treating signal C as clock input; (d) resolve hazard by inserting another vector between hazard-causing pair.

Thus in automatic test generation they will be treated as clock signals by activating them after other input vectors have been applied as shown in Figure 3.17c.

A less effective way to alleviate the problem is to design the ATVG such that only one input bit can change between adjacent vectors. However, the order in changing the single input bit can determine what final value is latched by the circuit. For instance, the correct order is to change signal A first while keeping signal B at its original value, then change signal B while keeping signal A as shown in Figure 3.17d. This is equivalent to inserting another vector between the original pair of vectors. However, if we reverse the order by changing signal B first and then signal A, then the incorrect value (i.e., 1) will be latched at the output Z of this circuit. This method is not as effective as the method of identifying equivalent clock signals because an ATVG does not usually contain the intelligence of knowing which signal should change first. It is also not a reliable practice to assume that the input with the smallest pin number should change first.

In summary, design engineers must not overlook the importance of conveying the clocking information to the test engineers so that an ATVG can generate vectors with the equivalent clock input being kept inactive until the circuit stabilizes. Sometimes a test engineer can use this information to manually modify the test vectors in the test program to take care of potential hazard problems. As we have shown in Figure 3.17d, a test engineer can insert another vector between a pair of vectors that causes a hazard.

3.5.3 Do Not Use Logic Redundancy

Logic redundancies are the portions of a circuit that do not contribute additional functions to the circuit. They exist in a circuit because proper minimizations in logic design were not performed. Since the introduction of logic redundancy does not change the functionality of the circuit, engineers may have the misconception that the extra circuitry does not cause testing problems. But in practice, circuits with redundancies most often have testability problems in terms of low fault coverage because an ATVG can not generate vectors to detect faults in the redundant logic. What makes it an even worse problem is that the presence of redundancies can make other faults undetectable. This fault-masking effect has further adverse impact on the testability. In this section, we will first present an example to show how faults can be masked by redundancies and then an easy method to detect the presence of redundancies.

Figure 3.18a shows a circuit with a redundant fault in the second input of the top AND gate.[13] This fault is not only undetectable, it also masks the S-A-1 fault of the first input of the same AND gate. Note that the S-A-0 fault of the first input of the lower AND gate (see Figure 3.18b) is detectable by the vector ($A = 1$, $B = 1$, and $C = 0$) when this redundant fault does not exist. But the same fault can not be detected by this vector if the redundant fault also exists. It turns out that another vector ($A = 0$, $B = 1$, and $C = 0$) can detect this fault in the presence of the redundant fault. However, this is generally beyond the capability of an ATVG because it uses a single-fault model in test generation. Also note that since a redundant fault is itself undetectable in testing, it can not be forestalled from masking other faults.

Some of the logic redundancies exist in the reconvergent fanout nodes in a circuit. The Test Counting[15] concept addressed in Chapter 2 can be used to detect the redundant logic faults.[16] The determination of the existence of redundancies along reconvergent fanout branches is based on a systematic approach to drive a set of sensitive one (1^+) and sensitive zero (0^+) through each branch from the stem of the fanout node. By way of the detection of the presence of logic value conflict during backward drive from the reconvergent gate of the fanout loop, it determines the existence of redundant logic.

As we recall from Chapter 2, a 1^+ (0^+) is defined as a logic value that, when applied to the input of a gate and complemented, will cause the output of the gate to also change to make this complementing observable. Thus, a stuck-at fault on an input signal but opposite to the sensitive value applied can make the fault detectable at the gate output. Therefore, in order to test a S-A-0 fault for a signal, a 1^+ must be assigned to it. Likewise, a 0^+ is assigned if a S-A-1 fault is to be tested.

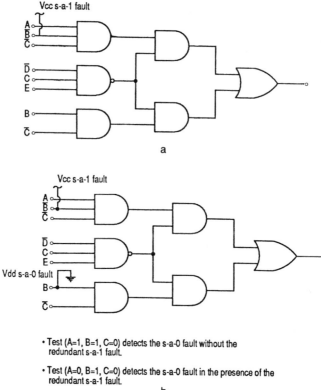

Figure 3.18 (a) Circuit with a redundant fault; (b) masking effect of the redundant fault.

We will illustrate this method using the circuit in Figure 3.19.[13] To determine if a redundancy exists that makes the S-A-0 fault undetectable on the upper branch of the loop (i.e., the first input to the top AND gate), a 1^+ is assigned to that branch while a 1^+ is also assigned to the other input of the same gate, as required by the rules in Chapter 2 to propagate a <u>sensitive one</u> through an AND

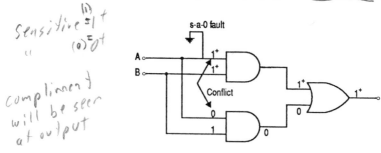

Figure 3.19 Test Counting method to determine the existence of redundant fault.

a gate. Note that to make the S-A-0 fault observable at the circuit output, a logic 0 is required at the other input of the OR gate. This in turn requires a 0 on at least one of the input signals of the lower AND gate, thus creating a conflict of logic values previously assigned to the other branch of the fanout node because all branches and the stem of the fanout node must have the same logic value in a given test. We can conclude that due to this conflict, a 1^+ is never allowed in the top branch of the reconvergent fanout loop and that the associated S-A-0 fault is not detectable. As we can see, this redundant circuit can be simplified easily by deleting the lower AND gate and thus eliminating the redundancy while preserving its functionality.

Following the same steps, we can determine that the S-A-1 fault in the circuit in Figure 3.18a is also undetectable because a 0^+ can not be assigned to the second input of the top AND gate without causing conflict. The associated sensitive values and logic values are shown in Figure 3.20.

In summary, redundant logic contributes to undetectable faults. Design engineers must understand the harmful effects of redundant logic on testing. The testing problems introduced by logic redundancies are not readily solvable. With redundancies in a circuit, an ATVG searches in vain for any paths to propagate the fault effect of the redundant faults and many backtracking steps can result. This can seriously impact its performance.

The best design for testability strategy is not to incorporate redundancies in circuit designs. If the existence of redundancies is unavoidable, then it is important for the design engineers to document such redundant logic so that test engineers can exclude the associated faults from the automatic test generation process to improve the efficiency of the ATVG used. Also in the fault coverage computation, these faults should be excluded because they are untestable as a result of design and not due to deficiencies in the ATVG or test procedures used by the test engineers. We will define a fault coverage called manufacturing fault coverage (MFC) as a true indicator of a circuit's testability constrained by its inherited DFT characteristics.

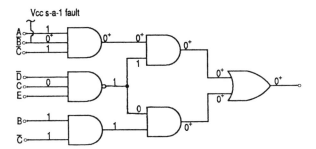

Figure 3.20 Another example of using Test Counting to determine the existence of redundant fault.

$$MFC = \frac{\text{detected faults}}{\text{total detectable faults}}$$

Note that MFC should not only exclude all the redundant faults but also faults due to the existence of illegal states in a sequential circuit. With both MFC and regular fault coverage computed for a design, the responsibility for testability improvements can be clearly delineated between the testing and the design engineering groups.

3.5.4 Avoid Designing Circuits with Illegal States

The design of a sequential circuit frequently starts with a state diagram based on the specification of the circuit's required behavior. To come up with a complete state diagram to satisfy the required behavior, the design engineer uses his or her experience and intuition as a guide. Once a complete state diagram is achieved, a decision is made on the type and number of flip-flops to be used for the design. The number of flip-flops (p) needed is related to the number of states (n) in the design by the following formula.

$$2^{p-1} < n \leq 2^p$$

Since the number of flip-flops needed can usually allow more states to be specified than in the design, there frequently exist unused states that can never be reached in normal operation. These unused states are called illegal states in testing. For example, the circuit represented by the state diagram in Figure 3.21 contains an illegal state when implemented as a sequential circuit using two flip-flops. It is also a noninitializable design. The testing problems for this circuit are that the ATVG has to rely on the information of its power-up state as the initial state to generate tests. However, if the power-up state of the flip-flops happens to be the "11" state, then this circuit is untestable because it is not possible to

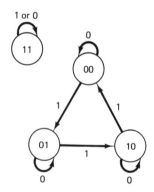

Figure 3.21 FSM design with an unused state.

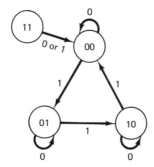

Figure 3.22   The FSM can reach other states from the unused state now.

reach other states from this state. On the other hand, even if the power-up state is one of the legitimate states, the circuit's testability will still be adversely affected by the existence of illegal states in the design. This is because there are faults that require the circuit to be in this state to be tested. Therefore, illegal states must be avoided to make the circuit testable. Faults associated with illegal states should also be excluded from the total testable faults in computing the manufacturing fault coverage defined in the last section.

The DFT consideration is to incorporate these unused states into a design and to allow transitions from any of these states to a common state, but disallow the transitions from any of the legitimate states into these states. For example, the FSM for the circuit in Figure 3.21 can be modified so that it can always reach the "00" state from the state "11" as shown in Figure 3.22, but it can not reach the "11" state from any of the other three states.

3.5.5 Add Extra Logic to Control Oscillation

A circuit with an asynchronous feedback loop can get into oscillation in response to some input sequence. Herein we define oscillation as a condition in a circuit that some of its signals repetitively change logic values without settling into a known stable state. This complicates the test generation problem because an ATVG requires a circuit to stablize after applying an input vector so that path sensitization and line value justification can be performed. Oscillation makes the behavior of a circuit difficult to predict because the circuit sometimes ends up in an unknown state even if the oscillation eventually subsides.

A simple example of an oscillating circuit is shown in Figure 3.23. We further assume that this circuit is designed to be an oscillator in the sense that the output signal Y will alternate between 0 and 1 if a 1 is applied to the input signal A. With a 0 applied to the input A, the output stabilizes in the value 1.

Extra logic can be added to the oscillator to disable the oscillating mode [17] and produce stable signals to allow automatic test generation. If the circuit is rede-

Figure 3.23 An oscillator circuit.

signed with an extra input signal B added to the NOR gate as shown in Figure
3.24, oscillation can be controlled. With ($A = 0$ and $B = 0$), the circuit output
stabilizes at $Y = 1$ and it stabilizes at $Y = 0$ with ($A = 0$ and $B = 1$). But if
$A = 1$ and $B = 0$, the circuit oscillates as before. With the extra logic, automatic test generation is possible because the stabilized output signal values for Y
can now be used for fault effect propagation as well as line justification. The
oscillator is thus avoided to prevent interference with the automatic test generation for other parts of the circuit.

   If a circuit were not designed to be an oscillator, but was found oscillating
during simulation, then it is still possible to redesign it to control oscillation.
Figure 3.25 shows two cross-coupled NOR gates which form a set–reset (SR)
flip-flop. If we apply the vector ($S = 1$ and $R = 1$), then the outputs of both
NOR gates assume a 0. We then apply the vector ($S = 0$ and $R = 0$), now the

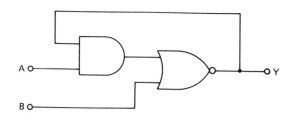

Figure 3.24 Same oscillator as in Figure 3.23 but with a control signal.

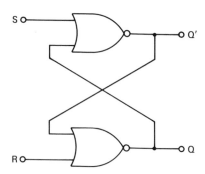

Figure 3.25 SR flip-flop formed by a pair of cross-coupled NOR gates.

NOR gates have a 1 at their output because both inputs to each NOR are 0. But the 1 at the output is fed back to the input of the NOR gate, which drives the output to a 0 again. This process is repetitive so that the output oscillates between 0 and 1 until a different vector is applied to inputs of the SR flip-flop.

Once oscillation is detected during simulation, the circuit can be redesigned to stop oscillation, and this can be done by adding an AND gate with three inputs (Q, Q', and A) as shown in Figure 3.26. The AND gate normally has an output signal 0 with its input signal A always tied to an X, where X indicates an unknown value. When we apply the oscillation-causing input sequence of vectors ($S = 1$ and $R = 1$) and ($S = 0$ and $R = 0$), the signal Y assumes a value X after the second vector is applied. As the X signal is fed back to the inputs of the NOR gates, their outputs take on the value of X because the other inputs (i.e., S and R) have the value of 0. This will stop oscillation in simulation. As far as automatic test generation is concerned, the X values can be used in fault effect propagation and line justification instead of a set of oscillating values of 0 and 1. Thus it does not interfere with the automatic test generation for faults in other parts of the circuit.

In summary, by adding extra logic the effect of oscillation in a circuit can be controlled so that automatic test generation is possible. However, the existence of an oscillator in a circuit should be communicated to the test engineer, including the oscillator's frequency and duty cycle so that proper arrangement can be made during testing to ensure that it works as specified. Communication between the design and test engineer about any other conditions that are potentially oscillation-causing is also important to facilitate automatic test generation.

3.5.6 Other DFT Considerations

The design for testability techniques discussed in the previous sections are effective to facilitate automatic test generation. We have emphasized these techniques

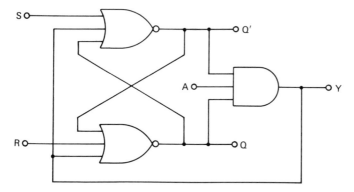

Figure 3.26 Redesigned SR flip-flop to control oscillation.

because automatic test generation has traditionally been the bottleneck in testing complex circuits. Even using these techniques, there may still be faults that are not detectable due to other factors such as sequential depth of the circuits and the fault models used by the ATVG.

In the remaining section of this chapter, we will discuss a structured design for testability method that actually converts a sequential circuit into a combinational circuit during automatic test generation and testing in order to circumvent the problems of testing complex sequential circuits.

There are also other DFT considerations important for actual testing of circuits on an ATE. We will postpone the discussion of them until Chapter 6 so that they can be discussed together with other ATE-related topics.

3.6   Scan-Path Design Techniques

In this section, we present a set of structured approaches to DFT that use some kind of scan techniques to access internal memory devices (i.e., flip-flops or latches) to facilitate the testing of sequential circuits. All these techniques [18] have two common characteristics as a result of being able to connect the memory devices into a serial path called scan-path during testing: (1) This allows states of a circuit to be shifted into these devices from one extra input pin; and (2) it also allows their contents to be shifted out from one extra output pin. This saves the need for a pair of input/output pins for each memory device. These techniques offer two distinct advantages in testing. The first is that the combinational logic and the sequential logic in a circuit can be separately tested. The other advantage is that a very simple method can be used to test the sequential logic without requiring the use of a modern ATVG to generate vectors for the sequential circuit. These methods have been used successfully in industry, especially by computer mainframe manufacturers. We will describe a popular scan-path technique called level-sensitive scan design (LSSD). But we will first describe a standard architecture commonly used in scan-path design.

3.6.1 Standard Scan-Path Architecture

Figure 3.27a shows a generalized architecture used by most of the scan-path techniques. The key differences between this model and the general sequential circuit model in Figure 3.2 is that there is an extra input pin S_{in}, an extra output pin S_o, and a test mode selection pin S_t all connected to the flip-flops. The flip-flops used in scan-path design must accept two input signals (see Figure 3.27b); D_1 is used for the data input during the normal mode of operation when $S_t = 0$ and D_2 is used for testing input when $S_t = 1$. Note that only one input signal is

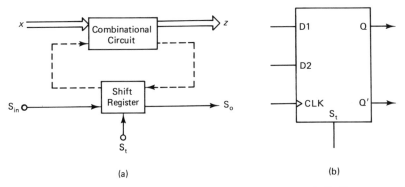

Figure 3.27 (a) Standard scan-path architecture; (b) a scan-path flip-flop.

accepted at a time by the flip-flop as determined by the external test mode selection input signal S_t.

3.6.2 Generalized Scan-Path Testing Procedures

Figure 3.28 shows such a structure for a circuit using two-input flip-flops to form a scan-path. During testing of the combinational logic portion of the circuit, all the flip-flops are again connected in series via the externally controlled test mode selection signal S_t to form a scan-path, with the output of one flip-flop connected to the $D2$ input of the next one to form a shift register. One of the flip-flops is also connected to the S_{in} pin so that a series of 0s and 1s can be shifted into these flip-flops by operating the clock signal. Once the flip-flops are set into the required states to detect faults in the combinational logic of the circuit, the circuit is reset to the normal mode of operation via the S_t signal for one clock period. During this clock period, the combinational logic can act on the flip-flop contents and the primary input signals. The test results are then stored in the flip-flops and can be shifted out through S_o after switching the circuit back to the test mode. Results of testing thus obtained can be compared with expected responses.

To accomplish testing of the combinational logic, the following steps are necessary:

1. Set $S_t = 1$, or the testing mode.
2. Shift the test values into the flip-flops.
3. Set the corresponding test value on the primary inputs.
4. Set $S_t = 0$, check the priamry output after the combinational logic settles.
5. Apply a clock signal and set $S_t = 1$ to shift out the flip-flops' contents via S_o to be compared with fault-free response values.
6. Shift the next test values into the flip-flops and repeat steps 3 through 5.

The sequential logic portion of the circuit is basically the scan-path flip-flops that must be tested first and separately from the combinational logic testing. This

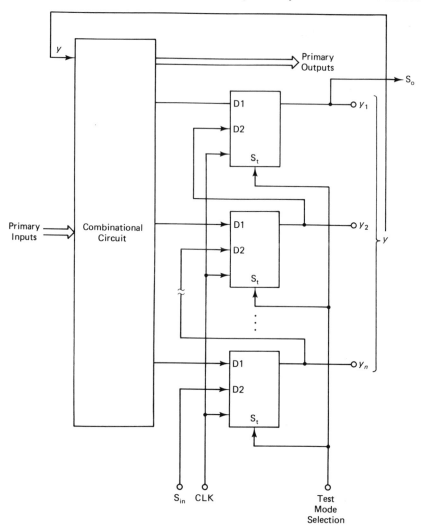

Figure 3.28 Generalized scan-path circuit.

is done by shifting into the serially connected flip-flops via S_{in} first a series of
0s, then a series of 1s, and also a series of 0s and 1s in combination. The series
of inputs are then shifted out of these flip-flops via the S_o pin to flush out any
faults in the scan-path.

We present an example to illustrate how scan-path techniques can facilitate
testing of a simple sequential circuit such as the one shown in Figure 3.29. The
circuit has two D-type scan-path flip-flops, a single primary input A, and a single

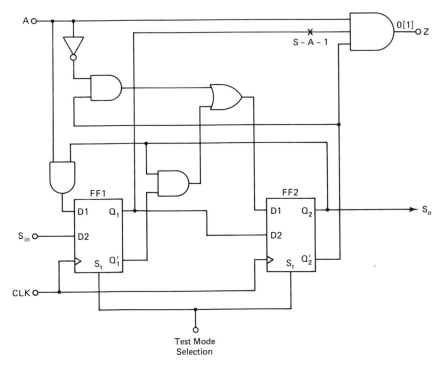

Figure 3.29   A scan-path example.

primary output Z. To test the stuck-at-1 fault for the second input signal of the right AND gate requires a signal value of 0 on this input and a signal value of 1 on the other two inputs. Note that the state needed is ($Q_1 = 0$ and $Q_2 = 0$) and also $A = 1$. Since this circuit does not have a master-clear signal, it is impossible to drive it into the 00 state through the primary input signal A only. Therefore this fault is not testable without using the scan-path design technique. With scan-path design, the 00 state can be easily shifted in through S_{in} following the steps outlined above, thus making the S-A-1 fault detectable. Furthermore, other faults such as the stuck-at-1 fault for the third input signal of the right AND gate is testable by shifting in the 11 state and by applying an input signal value for $A = 1$.

3.6.3 Level Sensitive Scan Design

The level-sensitive scan design (LSSD)[19,20] is one of the most successful scan-path techniques in use today. It uses special memory devices and design rules to reduce the number of dedicated input/output pins for testing and to eliminate most timing problems. The memory device—the shift register latch (SRL)—is

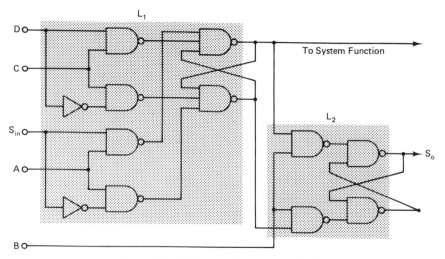

Figure 3.30 Shift register latches used in LSSD.

designed to be insensitive to such transient characteristics as rise time, fall time, and minimum circuit delay. The design does require that the clock remain high until the feedback loop stabilizes. Figure 3.30 shows two interconnected "level-sensitive" latches [18] which include a data input D, clock C, external scan-in input S_{in}, two nonoverlapping clocks A and B, and scan-out output S_o. Latches L_1 and L_2 function in a master/slave configuration during testing, while L_1 also functions as a storage device during the normal mode of operation.

Interconnecting several SRLs gives the circuit a scan capability (see Figure 3.31), which permits shifting any desired state in and out of the circuit. During

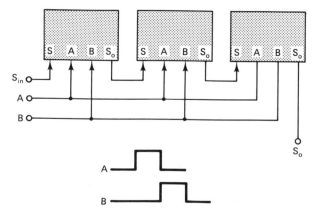

Figure 3.31 SRLs form a shift register in testing mode.

testing the scan-out output S_o for every SRL is connected to the scan-in input S_{in} of the next SRL and, by using two nonoverlapping clocks A and B, the circuit becomes a shift register.

Figure 3.32 shows an application of the LSSD structure[18] to a generalized circuit. Here, with the circuit in the normal mode of operation, it uses the system clocks C_1 and C_2 in a master/slave mode. During testing, a desired test vector for the combinational logic portion of the circuit is shifted into the SRLs via the shift clocks A and B. The results of the test applied to the combinational logic appear at the input of the L_1 latches. The circuit then compares this response with the expected response. To test the SRL, shift in and out a sequence of 0s and 1s.

The advantage of the approach just outlined is that it reduces the testing of a sequential circuit to the testing of a combinational circuit. This technique permits partitioning a large circuit into manageable segments of shift register–combinational circuit for which test vectors are easily generated. In addition, this approach does not affect normal system operation. In general, the LSSD tech-

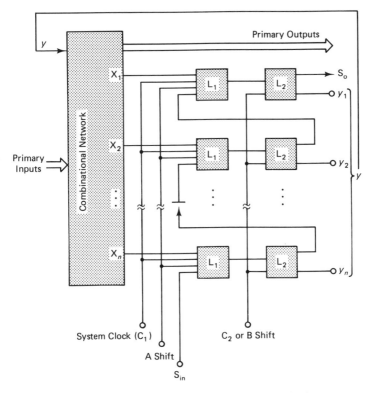

Figure 3.32 Application of LSSD in a generalized circuit.

nique eases the testing problem, but not without some cost in circuit performance and an increase in chip area overhead.

3.7 Summary

Automatic test generation has been considered one of the most critical steps in testing complex sequential circuits because of the difficulty in deriving test stimuli for high fault coverage. Modern ATVG techniques are frequently found inadequate even to handle a moderately complex circuit, especially with many levels of embedded memory devices. A sequence of test vectors instead of a single vector, as for the case of a combinational circuit, is needed to successfully test a specific fault, and these vectors must be applied in their proper order. For sequential circuits, there are also other problems such as the circuits' initializability and the existence of illegal states that can severely impact their testability.

In this chapter, we have addressed these problems from the standpoint of design for testability to facilitate automatic test generation. We also presented scan-path design techniques as a structured approach to DFT. But the ultimate responsibility of testability of a circuit rests on the design engineers and only they can solve the testing problems by remembering to incorporate testing considerations while doing their design.

References

1. Putzolu, G.R., and J.P. Roth, A heuristic algorithm for the testing of asynchronous circuits, *IEEE Transactions on Computers* **C-20,** June 1971.
2. Cheng, Wu-Tung, Split circuit model for test generation, *Proceedings of 25th ACM/IEEE Design Automation Conference,* June 1988.
3. Muth, P., A nine-valued circuit model for test generation, *IEEE Transactions on Computers* **C-25,** June 1976.
4. Marlett, Ralph, EBT: a comprehensive test generation techniques for highly sequential circuits, *Proceedings of IEEE Design Automation Conference,* June 1978.
5. Marlett, Ralph, An effective test generation system for sequential circuits, *Proceedings of ACM/IEEE Design Automation Conference,* June 1986.
6. Cheng, Wu-Tung, The back algorithm for sequential test generation, *IEEE International Conference on Computer Design,* October 1988.
7. Mallela, Sivanarayana and Shianling Wu, A sequential circuit test generation system, *Proceedings of 1985 International Test Conference,* October 1985.
8. Agrawal, Vishwani D., Kwang-Ting Cheng, and Prathima Agrawal, CONTEST: a concurrent test generator for sequential circuits, *Proceedings of ACM/IEEE Design Automation Conference,* June 1988.
9. Goldstein, L.H., Controllability/observability analysis of digital circuits, *IEEE Transactions on Circuits and Systems* **CAS-26,** September 1979.
10. Cheng, Wu-Tung, private communication.

11. Cheng, K.T., and V.D. Agrawal, A simulation-based directed-search method for test generation, *Proceedings of International Conference on Computer Design,* October 1987.

12. Miczo, Alexander, "Digital Logic Testing and Simulation." Wiley, New York, 1986.

13. Wang, Francis and Eric Engstrom, Designing PLD circuits for testability, *Electronic Design,* April 27, 1989.

14. Shiva, Sjjan G., "Introduction to Logic Design," Scott, Foresman and Company, Glenview, Illinois, 1988.

15. Akers, Sheldon B., and Balakrishnan Krishnamurty, Test Counting: a tool for VLSI testing, *IEEE Design & Test of Computers,* October 1989.

16. Hung, Angelo C., and Francis C. Wang, A method for test generation directly from testability analysis, *Proceedings of 1985 International Test Conference,* October 1985.

17. de Bruyn Kops, Peter, Testability Is Crucial In PLD-Circuit Design, *Electronic Design News,* August 18, 1988.

18. McCluskey, E.J., A survey of design for testability scan techniques, "Summer 1986 Semicustom Design Guide," *VLSI Systems Design,* CMP Publications, Manhasset, New York.

19. Eichelberger, E.B., and T.W. Williams, A logic design structure for LSI testability, *Proceedings of 14th Design Automation Conference,* June 1977.

20. Eichelberger, E.B., Latch design using level-sensitive scan design, *COMPCON 1983, San Francisco, California,* 1983.

Chapter 4

PLD Design for Test

4.1 Introduction

A programmable logic device (PLD) consists of an array of logic gates and flip-flops that can be programmed to implement a large number of logic designs. The engineer can specify a design using CAE tools which permit the design to be expressed with Boolean equations, a truth table, schematic diagrams, or state diagrams. PLDs give engineers an option between fixed-function devices (or "catalog logic") and custom integrated circuits. PLDs can be used to replace catalog logic if a product is being redesigned to save space. They are also excellent vehicles for prototyping custom IC designs. Because PLDs have become more and more powerful, many designs are implemented start-to-finish with PLDs. Programmable design can be done quickly and efficiently using CAE tools. If the PLD design needs to modified, it can also be done quickly and easily.

The need to design more compact digital circuits than is possible with SSI and MSI components—but without increasing design time and cost and hampering design flexibility—led to the popularity of PLDs. But PLDs also have unique problems in testing and design for testability because of their special architecture and the way they were manufactured. In this chapter, we address these problems from a generic point of view without referring to a particular type of PLD (e.g, PLA or PAL). The first question that was always asked is "why should PLDs be tested?". We will discuss the need for testing and the crucial role of testability in PLD design in the following sections.

The basic design and manufacturing cycle with a PLD is as follows.

- Conceive and describe the design
- Simulate and verify the design function
- Convert that description to the industry-standard JEDEC format

- Program the PLD
- Test the programmed device
- Incorporate tested devices into board-level designs
- Perform in-circuit testing of boards with programmed PLDs

Programming and verifying the device are the domain of the PLD programmer, and the choice of programmer can have a significant effect on both programmed device yields and finished product reliability. Programming, of course, is the first point of interest, and PLD programmers vary widely in the number and types of devices they can program, whether or not hardware adapters are needed to accommodate different types, programming speed, and the use of manufacturing-approved programming algorithms. These are typical concerns. But the choice of programmer also has ramifications in testing.

Fuse verification, the first step in any thorough PLD testing and verification process, is a standard postprogramming task performed automatically by all PLD programmers. The only distinction here between programmers is the speed at which fuse verification is performed. The typical PLD programmer performs fuse verification in 0.5 to 2 sec for a 28-pin device.

Test vectors are needed to test PLDs just like they are needed to test other circuits. Once test vectors are created they are loaded into the programmer and applied to the programmed devices. Test vectors in JEDEC format are usually loaded along with the programming data, though this is a convenience not a requirement. High-end programmers often hold as many as 9,999 test vectors; low-end programmers may hold as few as 200. Therefore, it is important to select a programmer with the capacity to hold complete test vector sets for the devices to be programmed. It is also important that the set of test vectors created (by any techniques to be discussed) be the smallest set possible that still provides maximum fault coverage.

4.2   Why PLDs Should Be Tested

So far we have assumed all programmed PLDs should be tested. But in reality, when testing programmable logic devices is discussed, the first question asked is "why test PLDs at all?" It arises because early PLDs were often used only for the collection of glue logic, and testing these functions in a rigorous way seemed to be overkill. In early uses, the most complicated application of PLDs was address decoding, which didn't demand consistent testing efforts.

Today, however, the picture has changed significantly.[1] PLDs in common use contain thousands of gates, feedback loops, macrocells, programmable internal circuitry, and bidirectional and registered inputs and outputs. These increasingly complex logic devices are no longer used for collection of random logic. Instead, a PLD is most often used to implement some portion of a design's most critical logic. In many cases, a contemporary PLD contains the same function that in the past required a full circuit board. There has never been any question about testing circuit boards before they are incorporated into end products. Such testing has long been the industry standard. It follows, then, that the new PLDs that are used like circuit boards must be tested as thoroughly. The problem is only exacerbated by the fact that modern designs may incorporate many PLDs on one board. In 1988, estimates placed the number of PLDs on more advanced circuit boards anywhere between 40 and 80.

This suggests the need for testing PLDs, but statistics prove the point. For example, heavy users of PLDs report that, in the absence of testing, typical failure rates after programming are 1%. If just 1% of the PLDs on a board fail, a board with 10 PLDs on it will fail 9.6% of the time. If the number of PLDs on the board is 20, the board failure rate climbs to 19.2%. A board with 50 PLDs will fail almost 50% of the time—resulting from a 1% failure rate among the programmed but untested PLDs on the board. Clearly, steps must be taken to reduce the number of programmed PLDs that fail.

The most effective way to achieve this reduction is testing. Testing of a PLD should occur as part of the manufacturing process and should continue through every stage of its journey to a finished product: after-programming, in-circuit, and in-system tests. And the more thorough the tests at each stage of the manufacturing process, the greater the end product yield rates. And the earlier an error is detected, the greater the cost savings.

The 1% failure rate for programmed PLDs is composed of the factors[2] shown in Figure 4.1. Human error such as mislabeling or mishandling of the device

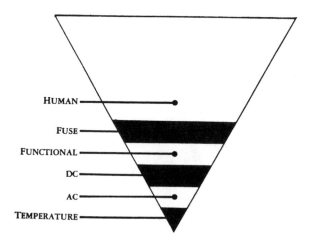

Figure 4.1 Factors that cause PLD faults.

before or after programming makes up about three-quarters of the 1% failure rate. With no testing, it's not uncommon for a device to make it onto a board without being programmed at all. Adequate training, improved manufacturing processes, and quality control are the only solutions to this problem. The remaining one-quarter of the 1% failure rate is due, in order of decreasing frequency, to device programming errors, functional errors, physical device faults, DC parametric faults, AC timing faults, and temperature-related failures.

DC parametric, AC timing, and temperature failures are virtually certain to be caught by manufacturing tests before the device ever leaves the semiconductor house. Therefore, these types of errors typically are of little interest to design and test engineers.

4.3 What Needs To Be Tested?

There are four types of possible errors or faults that can arise with respect to PLDs: parametric faults, design errors, programming errors, and device faults. We will describe each of these types of errors next.

4.3.1 Types of PLD Faults

Parametric faults are usually beyond the designer's control and involve either a process variation or an external circumstance that affect the performance of the PLD. Such errors may include slower than expected circuit speed and variances in current or voltage level caused by temperature, humidity, or manufacturing process. Generally, all PLDs delivered to a customer have been thoroughly tested for these types of faults, and the engineer can assume they won't occur.

Design errors result from the process of translating a logic design to the PLD implementation. Even when the highest level design languages and compilers are used to describe the PLD design, human errors can and do occur. Such errors may be as simple as using the wrong signal name in an equation or not accounting for an unexpected address in an address decoder. This is precisely why most high-level compilers provide simulation tools as well. A good simulator helps the design engineer detect and correct design errors before the actual PLD is ever programmed.

The detection of programming errors, design errors, and device faults, however, is up to the PLD user. Programming errors, which may occur as a result of either programmer or device failure, are relatively simple to detect via a process called fuse verification. This process checks to make sure that each fuse, switch, or transistor junction in the device was programmed as intended. Most commercially available programmers automatically perform fuse verification after programming is done.

While fuse verification is simple, fast, and automatic, it is not able to detect functional errors or device faults. For example, a functional error as programming a 16L8 design into a 16R8 device will pass a fuse verification test but will cause operational errors in circuit.

The detection of device faults such as shorts, opens, cross-point faults, double clocking, or powering up into unknown states is more difficult and requires the use of test vectors. To test a programmed PLD for all potential device faults, it is necessary to generate a comprehensive set of test vectors. Device testing vectors can be generated in one of many ways as we will discuss in the next section.

4.3.2 Methods of Test Vector Generation

To test a programmed PLD for all potential faults, test vectors can be generated in one of four ways: creating them manually; using exhaustive generation programs; using pseudorandom generation techniques; or using an automatic test vector generator (ATVG).

Though in theory any of the four methods can produce a set of test vectors that cover all potential device faults, all but ATVG have significant disadvantages.

Manual generation is tedious and introduces the possibility of human error. Exhaustive generation often creates more test vectors than the programmer or ATE can handle. For example, it creates 65,536 test vectors for a 16-input device. Pseudorandom techniques, though fast, do not take specific device architectures into account, relying on probability instead. In contrast, ATVG programs are rapidly gaining popularity since they are fully automated and can create optimized sets of test vectors for the specific device under test.

4.3.3 Fault Modeling for PLDs

Some of the materials in this and the next section are adapted from Data I/O Corporation's product descriptions.[3] PLDs normally have regular structures including AND arrays, OR arrays, input and output drivers and buffers, registers, and XOR gates. The data input pins are connected to a PLD's AND array through a set of fuses. These fuses are kept intact for an unused PLD. But during the programming step of PLD manufacturing phase, some of the fuses are blown and some are kept intact, as shown in Figure 4.2. The intact fuses provide connections for data input pins to the corresponding inputs of the AND gates. The blown fuses essentially disconnect the unused data input pins from other parts of the circuit. But if an AND gate is not used in a design, the entire row of fuses connected to the input terms is left intact.

Fuse faults are important for PLDs because they can occur after programming. Some fuses are supposed to be blown but instead they are intact or vice versa. One of the instances that can cause a blown fuse faulted intact is the existence

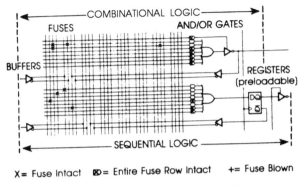

COMBINATIONAL LOGIC

FUSES   AND/OR GATES

BUFFERS

REGISTERS
(preloadable)

SEQUENTIAL LOGIC

X = Fuse Intact   ⊗ = Entire Fuse Row Intact   + = Fuse Blown

Figure 4.2   Fuse status of a programmed PLD.

of errors in the fuse addressing logic in a programmer that cause the wrong fuse to be blown. Another instance is the occurrence of shorts bypassing the fuse. These shorts can easily occur for bipolar PLDs during programming using a noncalibrated programmer that causes metal splatter. These fuse-related faults are generally represented by S-A-0 and S-A-1 faults. Figure 4.3 and 4.4 show, respectively, a S-A-0 fuse fault model and a S-A-1 fuse fault model[3] in a PLD as well as the corresponding test vector needed to test the fault.

Using the concept of equivalent faults, we can see from Figure 4.4 that all other stuck-at faults along the path from the output of the AND gate to the register's input and output, the tristate buffer's input and output are tested by the same vector that detects the fuse fault. Modeling every fault in a PLD is not necessary to effectively test and detect all the faults. The fault equivalence concept states that two or more faults are equivalent if they are detected by the same

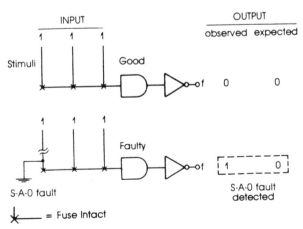

INPUT   OUTPUT

observed expected

Stimuli   Good

0   0

S-A-0 fault

Faulty

1   0

S-A-0 fault
detected

= Fuse Intact

Figure 4.3   S-A-0 fuse fault model.

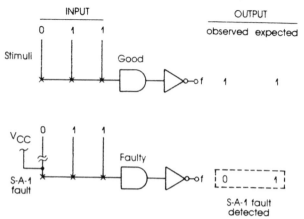

Figure 4.4 S-A-1 fuse fault model.

set of test vectors. This greatly reduces the number of fault conditions that have to be considered separately by an ATVG in generating test vectors for a PLD.

4.4 Effect of Device Types on Testability

The type of PLD can also affect its testability. In this section we will address the effect on testability of several types of devices, including erasable versus nonerasable devices and preloadable versus nonpreloadable devices.

4.4.1 Erasable versus Nonerasable Devices

Erasable programmable logic devices (EPLDs) present testing advantages to device manufacturers and design engineers. By definition, these devices can be erased at any time and reprogrammed with new data. For the semiconductor houses, this means preshipment tests can be performed that are the functional equivalent of fuse verification. The device can be reprogrammed and then checked against the programming data. Cycles of programming, erasing, and reprogramming can be performed for comprehensive testing.

Unlike erasable devices, one-time-programmable devices such as standard PLDs and PROMs present a one-time-only challenge; if the device operates incorrectly after programming, it must be discarded and another device must be programmed. Therefore, device manufacturers build special test circuitry into these devices so that they can be tested without programming the device. While these tests are highly effective at eliminating parametric and timing errors, they do not exercise the device in its operative state and thus do not provide any assurance that it will function correctly after programming. So it is doubly important to test one-time-programmable devices after programming through the application of carefully created test vectors.

4.4.2 Preloadable versus Nonpreloadable Devices

Preloadable PLDs have built-in functions to allow registers to be loaded from the output pins to any states consisting of 0s or 1s as required for testing. This capability aids testing of sequential circuits by direct setting of the device's output state. Typically a super-voltage normally between 10 to 13 V is applied to the output pins to force the registers into the desired states. Note that because of the output inverter, a register that has been preloaded High will provide a Low at the output.

Once all registers are set to known states, the test generation problems are similar to that of combinatorial circuits, making it unnecessary to cycle through long vector sequences to reach a desired state. Heuristic techniques[3] have been developed to allow efficient test generation for preloadable PLDs, and the test vectors thus generated normally achieve high fault coverage. A preloadable PLD allows all state transitions in a finite state machine (FSM) to be tested thoroughly because the FSM can be set to any initial states, including illegal states. Access to these illegal states make it possible to verify transitions from these states and to observe proper recovery. Because faults that would not normally be detectable can be detected now, higher fault coverage can be achieved compared to a nonpreloadable device. Another benefit as far as testing is concerned is that a minimum number of vectors are usually generated for preloadable PLDs.

4.4.3 Complex versus Simple PLDs

Complex PLDs can present more testability problems than simple ones. For instance, in a 22V10 device the clock signal can also serve as data input for other parts of the PLD and so can the asynchronous reset signal. Therefore, in testing of a 22V10 device the clock signal must be held high (i.e., use a nonreturn-to-zero mode) until the end of a particular test so that the input to the other parts of the circuit will not be changed in the middle of the test. To avoid registers to be reset to the zero states, the asynchronous reset input should not be activated during normal testing. The asynchronous reset input should be tested independently.

4.5 JEDEC Standard Format

JEDEC[4] format is today's most widely acceptable standard for field programmable and support tools. It defines a data format for transferring the fuse state from the development system to the programmers. The fuse state is specified in a graphical representation of the fuse state called fuse map. JEDEC standard also defines a simple testing format to be discussed in the following section.

4.5.1 JEDEC Test Vector Format

A test vector for testing a PLD is normally specified in the JEDEC format. It is simply a string of input values that is applied to the device inputs followed by a string of expected output values that is checked against the actual outputs obtained. Each test vector has one value specified for each device input pin and output pin. Therefore, a 20-pin device requires a 20-character vector. The test conditions contained in the fields are applied to the device pins in numerical order from left to right. That is, the leftmost field contains the test condition applied to pin 1, and the rightmost field contains the test condition applied to pin 20 for a 20-pin device. Valid characters for each field of the vector are as follows.

1	Drive the pin to a logic High
0	Drive the pin to a logic Low
H	Test the pin for a logic High
L	Test the pin for a logic Low
Z	Test the pin for a high-impedance state
C	Apply a positive going clock pulse
K	Apply a negative going clock pulse
X	"Don't care" state
N	No test performed
P	Execute a register preload
B	Execute a buried register preload

A typical test vector is shown below. Note that a positive going clock pulse is applied to pin 1; "don't care" conditions are applied to pin 2 through pin 4; the sequence 10110 is applied to pin 5 through pin 9, no test condition is applied to pin 10 and pin 20, a logic value 0 is applied to pin 11; and the sequence LHLLLHLH is expected at the output pin 12 through pin 19. Note that the * indicates the end of a vector.

CXXX10110N0LHLLLHLHN*

A typical vector sequence for a preloadable PLD normally starts with a preload vector to set the register state, then a vector verifying the output values of the preloaded registers, and finally a clocked vector to detect specific faults associated with the register state. As an example, a three-vector sequence is given below for a preloadable 16R4 device. The logic diagram for a 16R4 device is shown in Figure 4.5.[3]

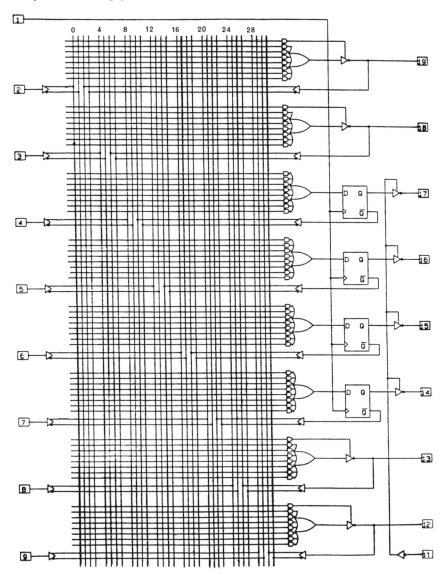

Figure 4.5 Logic diagram of a 16R4 device. (Courtesy of Data I/O Corporation.)

10 - GND
20 - VCC

V01   PXXXXXXXXXN1XX1101XXN∗   Preload
V02   0XXXXXXXXXN0XXHHLHXXN∗   Test (don't clock)
V03   C0101XXXXN0XXHLHLXXN∗   Next state

Note in vector V01, the first field has a character P to indicate that it is a preload vector. This is an exception in the correspondence between the test conditions and the pin numbers discussed above. The 0 and 1 input conditions should be used instead of the L and H output test conditions to specify the preload register values. Also note that pin 11, which controls the tristate outputs of the registers, is set to a logic 1 in the preload vector and that it is set to a logic 0 in subsequent vectors V02 and V03 when the normal output conditions are to be checked by a programmer or component tester.

Virtually all commercially available programmers have the ability to apply sequences of vectors to either a preloadable and nonpreloadable device. A programmer can stimulate the device inputs, check the device outputs, and report any discrepancies. This test usually follows programming and fuse verification. However, a preloadable device has a major problem for board-level testing. That is the incompatibility of preloading with in-circuit testing. The super-voltage required to jam load a register to a known state can damage other devices and parts such as a microprocessor connected to the PLD through a common bus on the same board. It may also impact the accuracy of parametric tests immediately after preloading.

4.5.2 JEDEC Fuse Map

Each fuse of a device is assigned a decimal number and has two possible states: a zero specifying a logical connection between two points or a one specifying a nonconnection between two points. The fuse number starts at zero and increases consecutively to the maximum fuse number. For example, a device with 2048 fuses has fuse numbers between 0 and 2047. All user programmable fuses can be specified with an L field. The L field starts with a decimal fuse number and is followed by a stream of fuse states (0 and 1).

The implementation of a design in a 16R8 device can show how the JEDEC fuse map relates to the actual PLD. We will represent a given design by its Boolean equation below.

$$!C = (A!C + AB) \tag{1}$$

Note that the exclamation point in front a variable means the complement of that variable. Figure 4.6 shows how this gets mapped into the 16R8 device. The logic diagram contains twice as many vertical lines (column) as inputs because there is an inverting and noninverting input from each pin connecting to its own vertical line. Therefore, the number of columns equals two times the number of inputs (also including I/O pins and feedback).

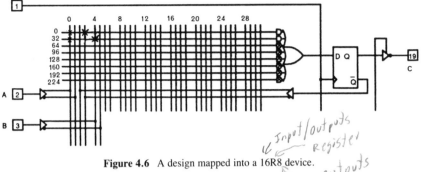

Figure 4.6 A design mapped into a 16R8 device.

In this example, the output *C* is connected to pin 19 and the inputs *A* and *B* connected to pins 2 and 3. The first term $A!C$ in equation (1) means that the first and fourth fuses in row 1 are left intact and all the rest in this row are blown. This leaves the noninverting input from pin 2 (fuse number 0) and the inverting feedback input $!Q$ from pin 19 (fuse number 3) connected to this product term. The second term AB in equation (1) means that the first fuse and the fifth fuse in row 2 are left intact while all the rest in the row are blown, which leaves the noninverting input from pin 2 (fuse number 32) and the noninverting input from pin 3 (fuse number 36) connected to this product term.

The corresponding JEDEC fuse map for the same design is shown in Figure 4.7. Note that the third through eighth rows contain all zero fuse state because these are fuses associated with the unused portion of the device and they should be left intact.

4.6 DFT Considerations for PLDs

The same degree of difficulty in test generation exists for nonpreloadable PLDs[3] as was discussed for sequential circuits in Chapter 3. It is caused primarily by

```
L000  011011111111111111111111111111*
L032  011101111111111111111111111111*
L064  000000000000000000000000000000*
L096  000000000000000000000000000000*
L128  000000000000000000000000000000*
L160  000000000000000000000000000000*
L192  000000000000000000000000000000*
L224  000000000000000000000000000000*
```

Figure 4.7 Corresponding Fuse Map for the Design of Fig. 4.6.

the lack of a direct means to initialize a registered PLD to a known state. One method is to generate a sequence of vectors to be applied to the input pins of the device under test so that signals can propagate through the AND and the OR planes to the registers to set them to specific states. When these registers are interconnected through feedback lines, the vector sequence can be long in order to set all registers to known states. This is due to the fact that for sequentially deep circuits or circuits with many levels of feedback connections, this process frequently requires setting one register at a time then finding additional vectors to propagate the known value to the rest of the registers.

Factors that affect testability of general sequential circuits discussed in Chapter 3 also affect PLDs. There are also special testability considerations for PLDs because of such factors as their unique structures, manners in which they are used, and methods of test program development. The time to consider ways of testing a circuit is before it is designed. This requires the design and test departments work together. The test engineer should participate in the design decision-making process and also provide guidelines for the design engineer about DFT considerations.

In the following sections, we will discuss DFT considerations for PLDs [3,5,6,7] in terms of some of the common design problems a device can have.

4.6.1 Eliminate Logic Redundancy

As was discussed in Chapter 3, faults associated with logic redundancy are not detectable and therefore reduce testability in a design. The end result for a design with redundancies is that the resultant circuit will have low fault coverage and the product incorporating the circuit may malfunction because of the existence of untested faults.

For PLDs, some of the logic redundancies can be found in the AND and the OR planes when reconvergent fanout exists. The existence of these reconvergent fanout nodes in a given design is frequently due to the lack of logic minimization or insufficient minimization in the design steps. These redundancies can be removed easily through logic minimization. For the circuit shown in Figure 4.8, the S-A-1 fault in the input A is undetectable due to the existence of logic redundancy. When the redundancy is removed from the design in Figure 4.8, the S-A-1 fault is detectable in the resultant circuit shown in Figure 4.9. Although manual logic minimization can be tedious and sometimes beyond human ability if a circuit uses many inputs, computers have automated these functions and today's logic design tools [8] usually can be used to perform logic reduction at various levels. A design engineer should never try to skip this necessary step in design and should always try to reduce potential logic redundancies to the maximum degree possible. Only by eliminating all extra product terms and extra literals in the remaining product terms will a completely testable design result.

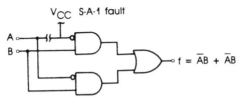

Redundant Circuits = Undetectable Faults

$$f = \overline{A}B + A\overline{B}$$

TRUTH TABLE (with redundancy)

| INPUTS | | OUTPUTS | |
A	B	f observed	f expected
0	0	0	0
0	1	1	1
1	0	0	0
1	1	0	0

Figure 4.8 A design with logic redundancy.

However, if redundant logic is required to control timing glitches, it can be tested by designing controlled-input test lines into the circuit. These test lines are active only during testing of the device but are held at a logic Low to allow the redundant logic to function during normal operation. This effectively removes the redundant logic when the test lines are active to permit the propagation of fault effects to the output pins where they can be detected.

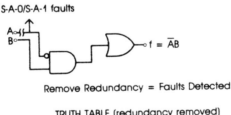

Remove Redundancy = Faults Detected

$$f = \overline{A}B$$

TRUTH TABLE (redundancy removed)

| | INPUTS | | OUTPUTS | |
	A	B	f observed	f expected
	0	0	0	0
	0	1	0	1
S-A-1	1	0	0	0
	1	1	0	0

S-A-1 fault detected

Figure 4.9 The same design as in Figure 4.8 but with logic redundancy removed.

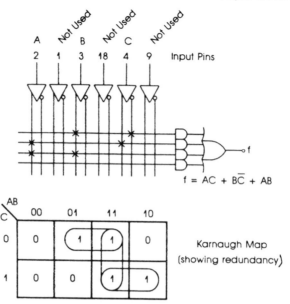

Figure 4.10 shows the input pins A (2), Not Used (1), B (3), Not Used (18), C (4), Not Used (9), feeding into buffers, a PLD array, and an output gate producing:

$$f = AC + B\bar{C} + AB$$

Karnaugh Map
(showing redundancy)

Figure 4.10 A PLD with redundant logic.

A design with redundancy is shown in the schematic diagram in Figure 4.10, and the associated Karnaugh Map indicates the redundant logic. Note that the addition of a test line increases the observability of faults at the output because it effectively eliminates the redundancy as shown in the schematic diagram in Figure 4.11. The associated Boolean equations illustrate the difference made by the addition of a test line to the logic function during testing.

4.6.2 Design Circuits That Can Be Initialized

Serious testing problems will occur for designs that can not be initialized from any arbitrary state into a known initialized state. For some PLDs the information regarding the device's power-up state is available from manufacturers. If so, then this information can be used to aid test generation and actual testing of noninitializable devices. But this scheme is not practical if the device has to be reinitialized many times in an AC or DC parametric testing where many vector sequences need to be run. Some design engineers count on the preload ability of a PLD to initialize the circuit, but this again is an impractical approach because preloading a device can lead to damaging other parts during in-circuit testing as discussed previously.

Initialization should be designed into a device. This can be done in a very simple manner at the cost of one extra input pin and one product term on some

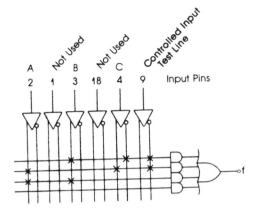

Test Line Held Low
(normal and testing operation) $f = AC + B\overline{C} + AB$
Test Line Set High (testing only) $f = AB$

Figure 4.11 Fault testing with additional control line.

outputs. The following steps show how to add initialization to an otherwise complete design described by Boolean equations.

1. Determine the state the device is to be initialized to.
2. Based on the initialized state, determine the corresponding logic value of each output pin.
3. If a corresponding output pin has a logic value Low, then add an initialization input I to every product term for that output.
4. If a corresponding output pin has a logic value High, then add one product term consisting solely of the initialization input I or its inversion input $!I$.

As an example, we will look at the noninitializable design shown in Figure 4.12. The Boolean equations to describe the design are given below:

$$Q1 = (!A)\,(!Q2)\,(!Q1) \tag{2}$$

$$Q2 = (!Q1)\,(B)\,(!Q2) + (!Q1)\,(A)\,(!Q2) \tag{3}$$

This design can be made initializable by following the four steps listed above. Assuming the state to be initialized to is 01 (i.e., $Q1 = 0$ and $Q2 = 1$), we will add the initialization input I to the product term in equation (2) and also add one

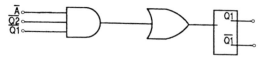

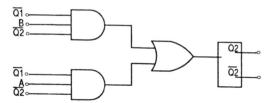

Figure 4.12 A noninitializable design.

extra product term consisting solely of the inversion of the input I to equation (3). The resulting equations are

$$Q1 = (!A) (!Q2) (!Q1) (I) \qquad (4)$$

$$Q2 = (!Q1) (B) (!Q2) + (!I) + (!Q1) (A) (!Q2) \qquad (5)$$

As can be observed from the new circuit diagram corresponding to the above two equations in Figure 4.13, initialization is accomplished by applying a logic 0 to the initialization input I and clocking once. This will set the $Q1$ output to a logic 0 and $Q2$ output to a logic 1. This example illustrates how straightforward it is to design an initializable circuit.

4.6.3 Provide Recovery Path from Illegal States

A finite state machine contains illegal states if some of the states or outputs are not specified for all the combinations of inputs and present states. When a FSM is implemented in a registered PLD, there are 2^n different states that can be assigned to a group of n registers in the device. If some of the states are not used in the FSM design, then they can be left as "don't cares" in the state table to simplify the logic and hardware realization. They are called illegal states because normally the device should not ever enter these states. However, due to transients or other unforeseen reasons the FSM can end up in one of these illegal states and have no way to return to one of the normal states. As a result, the device will have to be powered down and then restored to normal operation through a power-up and restart sequence.

One way to fix this problem is to provide a recovery path that will take the

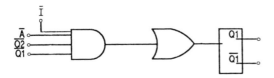

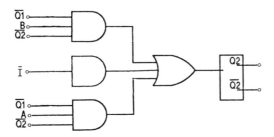

Figure 4.13   Initialization is designed into the circuit.

FSM back to one of its legal states.[9] Once this desired legal entry state is decided, then the "don't cares" in the Karnaugh Map for each of the state bits should be replaced with the corresponding bits value for the desired legal entry state.

An example of a six-state FSM implemented with a three-register circuit is shown in Figure 4.14. Since a maximum of eight states is possible with three registers, there are two illegal states in the FSM. We assume that the legal entry

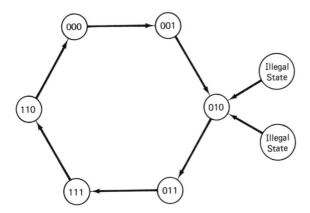

Figure 4.14   Recovery from illegal states into a fixed legal entry state.

state for recovery is state 010 (i.e., $Q1 = 0$, $Q2 = 1$, and $Q3 = 0$, where $Q1$ through $Q3$ are the three register values). The "don't cares" should be replaced with 0 in the Karnaugh Map for $Q1$, 1 in the Karnaugh Map for $Q2$, and 0 in the Karnaugh Map for $Q3$. Once these Karnaugh Maps are filled with the corresponding next state bit value, the equations are derived by including either all "don't cares" if filled with 1s or none of them if filled with 0s.

With a recovery path provided for every illegal state, testing is possible to verify the recovery procedures. Preloadable test vectors can be used to set the FSM into each of its illegal states and then to check for their transitions into the legal entry state. The faults associated with these illegal states can also be tested this way.

4.6.4 Avoid Latch Hazard

A latch can be formed easily in a PLD by using positive feedback in the combinatorial portion of the device. Therefore, a latch can be detected easily in a device if an output is a function of itself in the noninverting form. To be useful, a latch must be able to set or reset to a logic 1 or 0. A simple latch that has both a set and reset input is shown in Figure 4.15. Note that the output will stay at its present state of 0 or 1 until it is reset to the other state. For instance, if the set input is made High while the reset input is kept Low, then the output of the latch remains at a logic High. The output will go Low only if the reset input goes High.

A more general form of a latch involves several inputs on the set and the reset AND gates as shown in Figure 4.16. If the two inputs A and B change logic values in opposite directions from 0 to 1 and from 1 to 0, respectively, with a time interval during the transition that both inputs are 1, then glitches will occur that can cause a false latch at the output. Note that the output Y has a logic High instead of an expected logic Low.

To avoid this problem, edge-triggered flip-flops should be used to synchronize

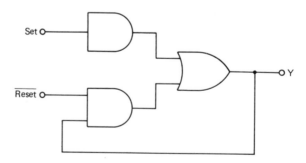

Figure 4.15 A simple latch with a set and a reset input.

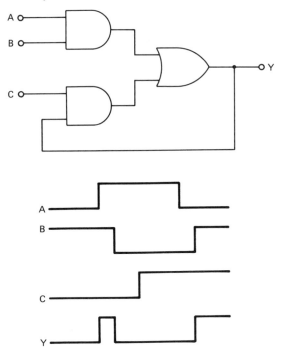

Figure 4.16 Hazard that causes a false latch.

the input signals so that the there is no time interval that both *A* and *B* input signals are 1. This can effectively eliminate these glitches.

4.6.5 Design Programmable Tristate Control Logic with Care

For some PLDs, the tristate output is controlled through a product term instead of through a dedicated input pin. If the product term also involves other input or feedback signals which can disable the tristate output, then these functions will not be observable at the output. This can impact testing in several ways as explained in the next paragraph.

First, testing some of the faults requires that these other signals be at logic values that happen to disable the tristate device, but then the fault effect can not propagate to the disabled output. This can nullify the test if the faults are also unobservable at other outputs. An example is shown in Figure 4.17. When the *B* input controls both the circuit logic and the tristate output, then any fault that requires *B* to be at a logic Low can never be tested at its output. Similarly, a register's output can also be made unobservable if its tristate control logic is a

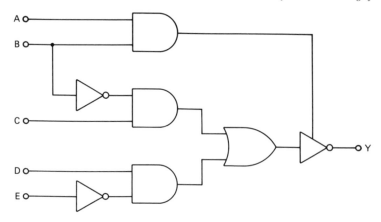

Figure 4.17 Untestable circuit with programmable tristate control logic.

function of its own value or of other signals. It is important during testing that test results always be observable.

The second problem in testing associated with programmable tristate control logic is its potentiality of causing race conditions in a device. The latch design shown in Figure 4.18 may require[6] both the tester and the device to drive the output simultaneously in order to latch data properly. This can cause good devices to fail if an input leakage test is performed, as is normally performed by a PLD tester when a pin is initially driven High or Low. On the other hand, if the driver on a tester is turned off as soon as the tristate output is enabled, it frequently results in a race condition on the tester which can cause failure of good devices. To circumvent this problem, the input A to control the tristate logic should be removed from the product term for the latch output. This makes the

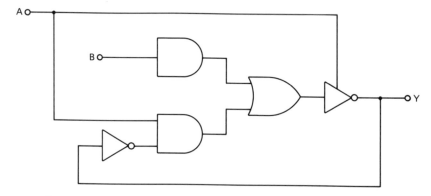

Figure 4.18 Another testability problem with programmable tristate control logic.

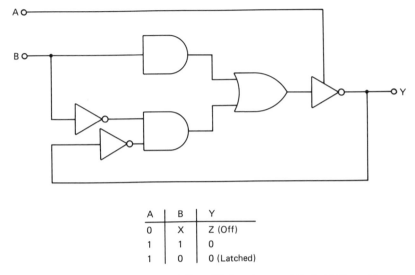

A	B	Y
0	X	Z (Off)
1	1	0
1	0	0 (Latched)

Figure 4.19 Design of Figure 4.18 modified to circumvent the testability problem.

tristate operation independent of the latch operation. Figure 4.19 shows the modified circuit where the input A has been removed from the feedback path. The corresponding latch operation is also included in the same figure.

4.6.6 Detect and Disable Oscillating Circuits

Circuits that oscillate can not be reliably tested because they keep on changing logic values without settling down to a stable condition. An oscillating circuit can occur accidentally and may not be detected during the simulation phase of design verification because an oscillatory circuit may be stable at times and may oscillate only under certain conditions.

An oscillator is formed when the circuit's outputs are fed back in the inverting form. A simple oscillating circuit that is untestable is shown in Figure 4.20. The output Y oscillates between 0 and 1 when the input A is holding to a logic value 1. If the circuit is modified by adding an enable input, then the output will stop

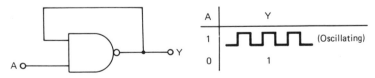

Figure 4.20 An untestable oscillating circuit.

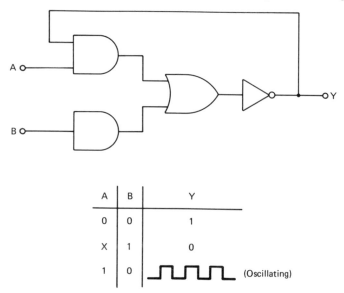

A	B	Y
0	0	1
X	1	0
1	0	(Oscillating)

Figure 4.21 An enable signal added to the circuit of Figure 4.20 to control oscillation.

oscillating when the enable signal goes High as shown in Figure 4.21. Now it is possible to test the circuit's output in its 0 and 1 state.

4.6.7 Satisfy In-Circuit Testing Constraints

There are additional constraints imposed on PLDs for in-circuit testing.[10] These constraints must be met in the design of PLDs in order to ensure successful testing of devices on a circuit board. When the devices driven by the PLD are tested, the in-circuit tester requires that the PLD be disabled (i.e., to set its output pins to a high-impedance state) to avoid level contention. Disabling a PLD in board-level testing can also avoid back driving other devices connected to its output into oscillation. Therefore, designing a PLD such that the output can be disabled is an important consideration for both sequential and combinatorial devices.

When a bus fault is detected, a test is required to determine which device on the bus is causing the fault. To generate the test, the tester needs to force all outputs of each device on the bus to a logic High and Low value. Some in-circuit board testers also require pins tied together during testing. For instance, two input pins are required to be tied together or a single pin to be tied to either a logic High or Low. These are all additional constraints for the design to conform, and they are also requirements for the ATVG used so that in-circuit testing can be performed successfully.

Another design for testability consideration for board-level testing is to avoid the use of external feedback from outputs to inputs of a PLD. External feedback connections can cause problems for most ATVG because this requires the interconnected output and input pins to be stabilized at the same logic value for each test vector generated. These feedback connections may also complicate in-circuit testing of the rest of the board. Since most PLDs have many internal feedback paths for connecting outputs to inputs, they should be used instead of external feedback connections.

4.7 Features Important for ATVG

As the applications of PLDs for prototyping and also in the final product increase, the devices become increasingly large and complex. Test engineers face the difficult task of choosing the ATVG most suitable for their PLD testing needs. What are the important features an ATVG must have in order to satisfy these needs? This is a difficult question to answer because each testing environment is different from the others. Features that are important for one company or one product may not necessarily be important for the others. We therefore can only summarize some general requirements and features of an ATVG for test engineers to consider in evaluating commercially available ATVG today.

4.7.1 High Fault Coverage

Fault coverage is an indicator of the ability of a set of test vectors generated to detect faults. High fault coverage means the thoroughness of a test, thus it ensures that device errors can be caught in testing. Therefore, high fault coverage is considered the most important requirement for an ATVG. In fault coverage analysis, care must be taken because the choice of faults to be included in its calculation can significantly affect the percentage fault detection number obtained.

The set of faults to be considered in the fault coverage analysis must include all fuse faults and cross-point faults. Here we define a cross-point fault as a fuse fault associated with an unused data input line. With reference to Figure 4.2, these are the fuses in the row that are marked with " + ." A fuse marked with a " + " is supposed to be blown. But for the reasons we discussed before, it can be faulted intact and therefore must be tested. All S-A-0 and S-A-1 cross-point faults should be included in the percent fault detection number. In another words, an ATVG must be able to generate vectors that not only detect high percentages of fuse faults but also cross-point fuses faults in order to ensure the thoroughness of testing.

However, faults associated with an entire row of fuses that are supposed to be intact (i.e., the associated AND gate is not used) need not be included in the computation of the percentage of fault coverage. This is due to the fact that normally all the data input lines to that row are tied to a logic Low so that the output of the AND gate connected to that row always has a logic 0 at its output. Unless all the data input lines to that row are stuck at a logic High value, the output will not change because individual data input lines stuck at a logic High value will have no impact on the AND gate's output.

One of the frequently asked questions about testing of any circuits is "how high a fault coverage should be achieved?" Based on the analysis of the impact of fault coverage on the board failure rate in Section 4.2, we suggest an objective of 95% or above to be used in evaluating an ATVG's fault detection capability. It is also important to verify the high fault coverage claim made by an ATVG for a given device on an independent fault simulator. Fuse faults that claimed to be detectable should be selectively injected into the fuse map which describes the circuit under test.

To inject a S-A-1 fault for an input line of an AND gate into the JEDEC fuse map, the zero (0) in the bit position corresponding to the input signal should be changed to a one (1). To inject a S-A-0 fault, a zero (0) should be placed in the bit position in the fuse map corresponding to the complement of the input signal, while leaving the original zero (0) for the input signal unchanged. Because of fault equivalence, other stuck-at faults in the OR plane, in the buffers, in the registers are equivalent to the faults in the AND plane of the PLD. To further verify the effectiveness of the vectors generated, a PLD should be programmed with these injected faults and tested on an ATE or on a programmer using the vectors derived originally for the unfaulted PLD to see if indeed these injected faults are detected.

4.7.2 Ability to Initialize PLDs

Why is the ability to initialize a PLD important? If an ATVG can not generate vectors to initialize a PLD, it must be tested using a known power-up state. This presents a formidable problem to the test engineer who has to perform in-circuit testing of PLDs on a board, because relying on the power-up state of a PLD for setting up its initial state requires power to be cycled before testing each PLD. This is impractical in a manufacturing testing environment where expensive automatic test equipment must be utilized to very high efficiency.

There are other reasons that an ATVG should be able to initialize PLDs. For instance, if the generated vectors have to be truncated due to vector length limitations of a particular device programmer or ATE, then the fault coverage will suffer. To prevent this, a test can be generated using vector sequences preceded by the initialization vectors. A test engineer can thus cut the test after one sequence, then load the remaining sequences to finish the test process.

It is also important to the test production engineers to know if a circuit is initializable because it will determine what type of PLD is ultimately selected.

For a noninitializable circuit, using a PLD that powers up to a known state will have better testability than one that powers up to an unknown state. This will have a direct impact on the options a purchasing agent has when procuring devices.

An ATVG must be able to initialize a PLD that is designed to be initializable from any initial register states. These initial states should include Unknown, Low, High, or as any combination of High and Low logic values for the registers in a PLD. This allows support of all possible power-up states of a device or facilitation of the generation of test vectors to be merged with other vector sequences, such as vectors provided by the design engineer to verify the functionality of the circuit.

4.7.3 Compatibility with ATE

Vectors generated by an ATVG for a specific type of test must conform to the constraints imposed by common ATE that performs the type of test. Certain ATE is capable of driving a bidirectional pin and monitoring it at the same time. If the ATE and the device under test drive to opposite logic values, then a conflict occurs.

The outcome of the resolution depends on the drive capability of the PLD or the ATE involved. In this case, either the tester will win, the device will win, or the outcome will be unknown. An ATVG must provide user-specified options for all three cases to resolve such a conflict so that a test engineer can choose among the three options. If the tester wins option is not chosen, then the test will fail when the device out drives the tester to an opposite logic value. For ATE that can not monitor while driving, the outcome of the conflict will be unknown and the outcome unknown option will have to be chosen.

In-circuit testing that detects pin faults due to manufacturing process requires certain pins tied together or tied to particular logic values. An ATVG must be able to generate test vectors when tied pins are specified in one of the four configurations: pins tied High (to V_{cc}), pins tied Low (to ground), input pins tied together, and pins not driven by the tester. Input pins tied to output pins is not a requirement for an ATVG because most PLDs provide feedback connections from an output to an input by way of its internal feedback paths.

Another testing requirement on an ATE is to measure the current driving capability of a device by connecting the output pin to ground through a resistor of low value while maintaining a logic High value on the pin. In this case, the vector generated by an ATVG must be able to maintain the output pin under test at a logic High state even when the output is drawn to ground.

4.7.4 Support Industrial Standard Formats

Use of industrial standard formats by an ATVG for output of test vectors and for accepting input data ensures easy interface with other CAE and CAT environments. Design and test of PLDs are always part of the overall activities for engineering and manufacturing of a product, no matter whether it is a computer or an electronic instrument. Therefore, modern ATVGs must include functions compatible with the front-end CAE tools for logic design and synthesis and back-end CAT tools to link up with functional and in-circuit testers.

All the characters that make up a JEDEC vector must be supported by an ATVG. This should include the Z value for high-impedance output, the C character for return-to-zero clocking mode, and the K value for return-to-one clocking mode.

On the other hand, logic design and synthesis tools are widely used by design engineers today. These tools usually include logic simulation, fault simulation, and logic minimization ability so that design verification and logic redundancy elimination can be performed. Therefore, the effectiveness of an ATVG in generating high fault coverage test vectors is also determined by its interfacing capability with these tools. The linking of an ATVG with a design and synthesis tool incorporating DFT considerations is the best assurance to produce testable designs. All such tools can accept several standard formats. An ATVG must support these formats in order to be compatible with these tools, thus taking advantage of the DFT features provided. For instance, an ATVG must be able to use the vectors developed by the design engineer to verify the design as seed vectors and generate additional test vectors to supplement these vectors in order to achieve high fault coverage in testing.

4.7.5 Provide Comprehensive Testability Report

As we all know, not all designs are testable. In order to improve a design's testability, an ATVG must provide data regarding the areas in a design that lack design-for-testability considerations. This information is useful for both the test engineer and design engineer involved. If problems are discovered early, then the design engineer can make the necessary changes based on the information provided. On the other hand, if the testability problems are discovered late in the manufacturing cycle, then the testing department will either reject the circuit all together or find ways to work around the testability problems.

Specifically, a testability report provided by an ATVG should include the following items.

1. Percent fault detection numbers for user-provided design verification vectors and any problems in simulating these vectors
2. Fault coverage statistics including percentage of faults detected and the number of total faults considered versus the number of faults detected
3. Initialization vector sequence or any initialization problems detected
4. Complete vector sequence and corresponding faults for each vector generated by fuse numbers and the associated pin numbers

5. Unreachable states in a finite state machine and the associated undetected faults

6. Boolean equations in terms of input and output pins that describe a design and the information of undetected faults associated with these equations

7. The existence of logic redundancies and the associated undetected faults

8. Run time statistics including CPU time, real time, and memory usage

9. Any other testability problems encountered such as race, timing hazard, and oscillation

4.7.6 Support Many Devices

For an ATVG to be useful, it must support the devices that a customer uses. If it supports a large variety of PLDs from different manufacturers, then a user will have the freedom to switch to other devices without having to deal with the problem of incompatibility with the ATVG used.

Since the architecture and the processing technology of PLDs are changing rapidly, new PLDs are constantly becoming available today. As a result, the time necessary for an ATVG to be updated and upgraded for new devices and new architectures must be considered before committing to any ATVG.

References

1. "1989 High Performance Systems Programmable Logic Guide." High Performance Systems, 1989.

2. Durwood, Brian and Francis Wang, How testing can save many headaches for PLD users, *in* "1989 High Performance Systems Programmable Logic Guide." High Performance Systems, 1989.

3. "PLDtest Plus User Manual." Data I/O Corporation, Redmond, Washington, March 1990.

4. JEDEC Standard No. 3-A, Electronic Industries Association, Washington, D.C., May 1986.

5. "PAL Device Data Book, Bipolar and CMOS." Advanced Micro Devices, Sunnyvale, California, 1990.

6. de Bruyn Kops, Peter, Testability is crucial in PLD-circuit design. Electronic Design News, August 18, 1988.

7. Wang, Francis and Eric Engstrom, Designing PLD Circuit For Testability, *Electronic Design,* April 27, 1989.

8. "ABEL User's Guide." Data I/O Corporation, Redmond, Washington, January 1989.

9. Shiva, Sajjan G., "Introduction to Logic Design." Scott, Foresman and Company, Glenview, Illinois, 1988.

10. "GR2282 Board Test System, Test Library Programming Guide." GenRad, Inc., Concord, Massachusetts, 1988.

Chapter 5

Built-In Self Test and
Boundary Scan Techniques

As we have discussed in previous chapters, the most difficult problem with testing is to generate test vectors. Test generation belongs to the class of the so-called NP-complete problems, and, in the worst case the time[1] required would grow exponentially with the size of the circuit. For large and complex circuits, the cost of product testing and its turn-around times can become prohibitive. The testing problem is compounded for some cases when less than satisfactory fault coverage is accepted because of the difficulty in test generation.[2] This prompted the search for more cost-effective alternatives than traditional test generation methods. One of the more economical alternatives developed is to have additional circuitry built into a digital circuit with automatic fault-detection capabilities. This leads to the birth of built-in self test (BIST), which is considered as a viable alternative to automatic test vector generation.

In the next few sections of this chapter, we will look at popular BIST methods that have been developed, discuss details required for implementing a particular popular BIST method, and offer guidelines in designing the built-in circuitry. We will then look at boundary scan, which is a design-for-testability technique for board level and higher levels of assembly of digital circuits and systems. It requires additional circuitry (register cells) placed adjacent to each I/O pin of an IC so that a scan chain can be formed by serially connecting all the register cells on the board. Thus test data can be shifted in and test results can be shifted out. Boundary scan techniques and their applications will be discussed in the second half of this chapter.

5.1 BIST Overview

Built-in self test is a method of testing digital circuits in which both test generation and test verification are done by circuitry built into the chip itself. Frequently, it uses a pseudorandom number generator constructed from shift registers as a source for test vectors. These vectors are applied to the circuit under test (CUT) at the circuit's clock rate. The test responses are then analyzed to determine if faults exist in the CUT. This is done first by compressing the raw response data into a signature. A precomputed signature for the fault-free circuit is used to compare against the signature computed from the actual response data to determine the pass or fail status of the CUT. A generalized BIST approach[3] is illustrated in Figure 5.1 in a block diagram. It consists of a functional block to generate test vectors to be applied to the CUT, a data compressor to reduce test response data to a compact signature, and the logic to compare the actual signature derived from the test data with a reference signature. All the testing control functions such as start of test, detection of the end of testing, and synchronization are under the control of the self-test controller.

The circuit under test can be either a combinational or a sequential circuit whose initial state is known before the test vectors are applied. Then depending the chip organization, various methods can be used to minimize the necessary testing circuitry.

5.1.1 General Characteristics

BIST in the most general sense can have any of the following characteristics.

1. Concurrent or nonconcurrent operations
2. Exhaustive or nonexhaustive test designs
3. Deterministic or pseudorandom generation of test vectors
4. Any one of the data compression techniques mentioned later in this section

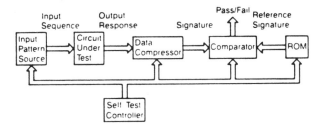

Figure 5.1 A general built-in self test approach.

Concurrent testing[4] is an on-line verification to check that portions of the computing process are being conducted correctly. It is designed to detect faults during normal circuit operation. *Nonconcurrent testing* is an off-line verification that requires that normal operation be suspended during testing. Therefore, it does not directly affect the reliability of system operation. Frequently, BIST is used for nonconcurrent testing.

An *exhaustive test* of a circuit requires that every intended state of the circuit be shown to exist and that all state transitions be demonstrated. For large sequential circuits, this is not practical, but for purely combinational circuits, an exhaustive test is the same as stimulating the circuit with the set of all possible input vectors.

When exhaustive testing is impractical, *nonexhaustive testing* is usually done. This generally involves the following.

* Modeling physical defects as faults
* Identifying all possible faults
* Developing a circuit description from which faults can be modeled
* Selecting test vectors to detect faults at the outputs
* Evaluating the resulting fault coverage

Deterministic testing occurs when specific modeled faults are targeted by specifically produced vectors. The vectors may require a ROM for storage, but they offer excellent fault location capability. *Pseudorandom testing*[5] occurs when randomlike test vectors are produced. The popularity of this approach is partly due to the fact that input vectors are easily generated both in hardware or software. In addition, shift registers that commonly exist in a circuit can be converted to linear feedback shift registers to generate pseudorandom test vectors during testing while performing their original functionality during normal operation. This reduces chip areas needed for the built-in testing circuitry.

Several techniques are also available to compress the response data into compact forms so that a pass–fail evaluation of the test response can be made easily. *Transition counting*[6] is one of the first compression techniques implemented in commercial trouble-shooting instruments. It basically counts the number of logic signal transitions in a binary response data stream and uses this number as a signature for the CUT. *Syndrome testing*[7] uses the ratio of the number of 1s that are present in the output to the number of possible input vectors as a signature.

The *signature analysis*[8] technique employs a linear feedback shift register (LFSR) to compress a response data stream. This approach can be interpreted as a form of polynomial division[4] and therefore is more amenable to formal analysis than any of the other data compression techniques. The design of a signature analyzer can also make this technique more error sensitive than other techniques.

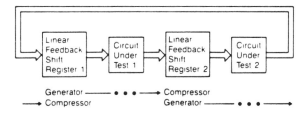

Figure 5.2 A BIST circuit using multipurpose LFSRs.

5.1.2 Pseudorandom Testing with Signature Analysis

When pseudorandom testing is combined with signature analysis, certain hardware efficiency can be realized by utilizing special multipurpose LFSR architectures. In addition, when these LFSRs are configured as memory elements of the functional design, even more hardware efficiency is obtained, along with an improved level of controllability, observability, and testability of the built-in self test circuit.

Most of the materials used in this section and the next section are adapted from Ref. 3. The basic approach for this BIST method is to use linear feedback shift registers both to generate pseudorandom test vectors for the circuit under test and also to efficiently compress its test responses to determine pass or failure status. A BIST implementation using LFSRs is shown in Figure 5.2. Note that the LFSRs can be built to selectively function as either a test vector generatoror as a test data compressor. As in this BIST arrangement, LFSR 1 generates test vectors for CUT 1 and compresses test data for CUT 2. On the other hand, LFSR 2 compresses test data coming from CUT 1 and generates test vectors for CUT 2. Another possible configurations for BIST implementation is shown in Figure 5.3 which has one LFSR to generate tests for two CUTs and one LFSR to compress data for two CUTs also. Other configurations of LFSRs are also possible.

5.1.3 Linear Feedback Shift Registers

A *linear feedback shift register* is a finite state machine made from three basic building blocks: adders, a memory device, and constant multipliers. Their sym-

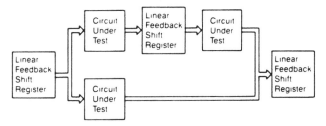

Figure 5.3 Another BIST configuration.

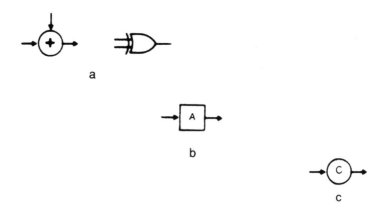

a

b

c

Figure 5.4 Components of an LFSR: (a) Adders; (b) memory device; (c) constant multipliers.

bols are shown in Figure 5.4. An adder is also called a modulo-2 summer. They are exclusive OR (XOR) gates. A memory device is also called a unit delay element. They are edge-triggered flip-flops. The value stored in the flip-flop (either 0 or 1) is referred to as the state of the memory device. Constant multipliers are used to make feedback connections. C has a value of either 0 or 1, where 1 indicates a connection and 0 indicates the absence of a connection.

There are two basic feedback configurations that are typically used to form an LFSR in BIST: internal exclusive OR (IE) type and external exclusive OR (EE) type. These are shown in Figures 5.5 and 5.6, respectively. The IE type is so named because the adders are embedded between the memory devices, while the EE type has the adders external to the forward signal path between memory devices. At any instant, the state of the LFSR is given by the values of the

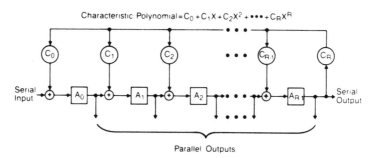

Figure 5.5 An internal XOR-type LFSR.

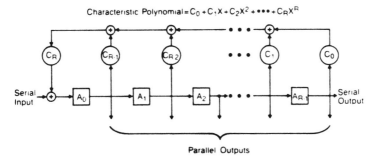

Figure 5.6 An external XOR-type LFSR.

memory devices. Note that any LFSR has a unique characteristic polynomial associated with it, where the coefficients of the characteristic polynomial are given by the feedback connections. For the same characteristic polynomial, both the IE and EE types of LFSR can be implemented. Note, however, that the relationship of the coefficients to the actual feedback connections are opposite between the two types (i.e., C_0 indicates the leftmost feedback connection in an IE type, but it indicates the rightmost feedback connection in the EE type).

Figures 5.5 and 5.6 represent the single-input versions of an LFSR, but, as will be shown, both types are easily adapted to accommodate multiple inputs for data compression. Note the reverse ordering of the feedback connections between the two equivalent configurations. These feedback connections are taken from polynomials called characteristic polynomials of the form

$$C_0 + C_1X + C_2X^2 + \cdots + C_nX^n$$

where C_i for i ranging from 0 through n is either a 0 or a 1. Selecting an appropriate characteristic polynomial is an important part of the design process and requires special attention, as will be discussed later.

The combination of values of all the memory devices on an LFSR defines its state. An LFSR with n flip-flops can assume any one of the $(2^n - 1)$ different internal states. The states that actually occur depend upon the following

- The initial state of the LFSR before the application of an input sequence
- The input sequence
- Feedback connections

With this background information about linear feedback shift registers, we will discuss design considerations of an LFSR for test vector generation and test data compression. Design objectives for a BIST circuit will also be highlighted.

5.2   Test Generation in BIST

We are primarily interested in the type of built-in self tests that are nonconcurrent, pseudorandom, and use signature analysis. Since most of this type of testing is nonexhaustive, the randomlike behavior is advantageous in that it allows analytical techniques to be used in estimating fault coverage.[9,10] It also gives an advantage over say, a binary counter, since it prevents masking of certain regular interval errors.

5.2.1 Test Vector Generation Using LFSR

To generate test vectors for the CUT, an LFSR is generally used without inputs. This is equivalent to a constant input sequence of zeros. The sequence of states then generated is dependent only on the initial state and the feedback connections. The test vectors generated are pseudorandom in nature because the vectors are equally likely and the same vector sequence can be repeated.

When using an n-bit LFSR, it is possible to generate up to $(2^n - 1)$ different states before repeating the sequence. A generator that repeats after exactly $(2^n - 1)$ states is called a maximal-length generator. To design an LFSR that is a maximal-length generator of test vectors is a prime design objective of a BIST circuit. A characteristic polynomial with coefficients C_i that yields a maximal-length generator is called a primitive polynomial. Thus, finding a suitable primitive polynomial is the first step in designing a maximal-length generator.

Now let's look at polynomials in general. All polynomials can be classified according to the tree structure shown in Figure 5.7. Irreducible polynomials can be classified into primitive[11] and nonprimitive polynomials. Note that a primitive polynomial must be irreducible (i.e., it can not be factorable into the product of two polynomials). As we discussed before, a primitive polynomial has the virtue that connections made from it result in a very useful property of LFSR. Specifically, as we clock through the states of the LFSR, we generate a maximal-length

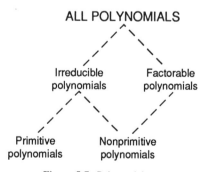

Figure 5.7 Polynomial tree.

n	2^n	Number of Primitive Polynomials
1	1	1
2	4	1
3	8	2
4	16	2
5	32	6
6	64	6
7	128	18
8	256	18
9	512	48
10	1024	60
11	2048	176
⋮	⋮	
20	1,048,576	24,000
⋮	⋮	
30	$(1.0737)(10^9)$	17,820,000
⋮	⋮	

Figure 5.8 Number of primitive polynomials.

sequence of vectors. While cycles produced by nonprimitive polynomials suffer from the following defects.

- Shortened cycle length
- Seed dependence
- Statistical characteristic change

Figure 5.8 shows the number of possible primitive polynomials for different values of n, where n is the number of bits in the LFSR. These can be obtained readily from standard mathematical books.

Except for the special case of zero initial state and constant zero inputs, an LFSR made from a primitive polynomial always generates a maximal-length sequence of test vectors. The initial state only determines where the sequence starts. Now let's take a look at a real example for the LFSR shown in Figure 5.9, where the feedback connections come from the primitive polynomial below.

$$X^4 + X + 1$$

From the same initial state (0001), note that both types of LFSRs generate a maximum sequence of 15 states, excluding the all zero state. Note the difference in the actual sequences of vectors generated by these two LFSRs with four memory devices each. Note also that the repetition starts after all 15 states have been cycled through.

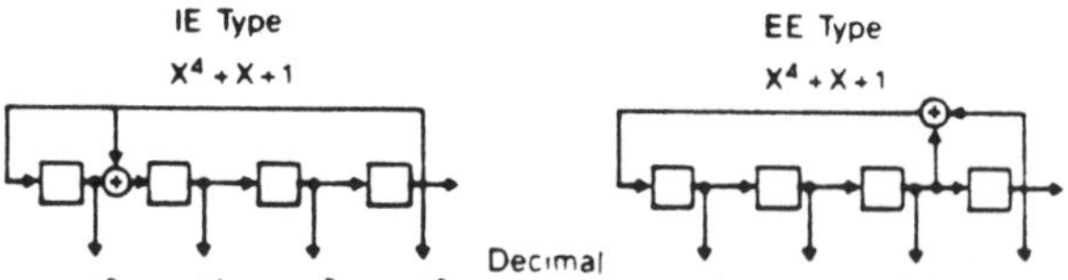

| | IE Type $X^4 + X + 1$ | | | | | EE Type $X^4 + X + 1$ | | | | |
Clock	A^0	A^1	A^2	A^3	Decimal Value	A^0	A^1	A^2	A^3	Decimal Value
Initial State	0	0	0	1	—	0	0	0	1	—
1	1	1	0	0	12	1	0	0	0	8
2	0	1	1	0	6	0	1	0	0	4
3	0	0	1	1	3	0	0	1	0	2
4	1	1	0	1	13	1	0	0	1	9
5	1	0	1	0	10	1	1	0	0	12
6	0	1	0	1	5	0	1	1	0	6
7	1	1	1	0	14	1	0	1	1	11
8	0	1	1	1	7	0	1	0	1	5
9	1	1	1	1	15	1	0	1	0	10
10	1	0	1	1	11	1	1	0	1	13
11	1	0	0	1	9	1	1	1	0	14
12	1	0	0	0	8	1	1	1	1	15
13	0	1	0	0	4	0	1	1	1	7
14	0	0	1	0	2	0	0	1	1	3
15	0	0	0	1	1	0	0	0	1	1
16	1	1	0	0	12	1	0	0	0	8

Figure 5.9 Maximal-length test generators.

When using coefficients of a nonprimitive polynomial for feedback connections in an LFSR, maximal-length sequences can never be generated. Moreover, the initial state can both influence the sequence length and determine which states occur. As an example, Figure 5.10 shows a test vector generator using the nonprimitive polynomial below.

$$X^4 + X^2 + 1$$

By observation, this polynomial is nonprimitive because it can be factored into the product of two polynomials. Note that only 6 out of a maximum of 15 states are generated before the sequence repeats itself. Another nonprimitive polynomial is shown in Figure 5.11, where 5 out of 15 states are generated before the cycle repeats itself.

Due to the limitations of the length of vector sequences generated, testing will not be as thorough as in the case of using a maximal-length test generator. Nonprimitive polynomials are not desirable for use as LFSRs in a BIST circuit because of this shortcoming.

5.2.2 Fault Coverage Analysis

The bottom line in test vector generation is that a usable, effective fault coverage must be chosen, and toward this end the classical single-stuck-at fault model is most frequently used for built-in self test. Historically, test vector sets that cover these faults have proven to be effective for detecting most of the faults that can

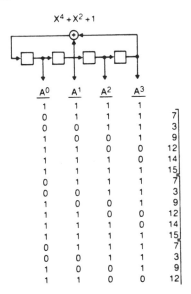

Figure 5.10   Test generator based on a nonprimitive polynomial.

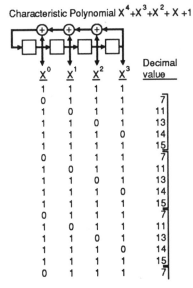

Figure 5.11   Another test generator based on a nonprimitive polynomial.

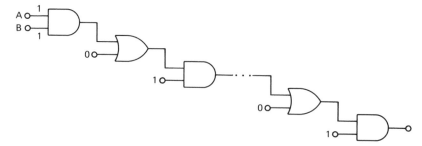

Figure 5.12 A cascading circuit with random-vector resistant faults.

occur in practice. In particular, they cover commonly occurring defects such as opens and shorts within gates.

Note that random vectors are not very effective in detecting typical failure modes in RAMs. Pattern-sensitive tests would be best for RAMs. There are also random-vector resistant faults within circuits with fan-in gates,[12] with logic redundancies, or with deeply leveled logic such as the cascade circuit in Figure 5.12. To detect the S-A-0 fault on input lead *A* of this AND–OR cascading circuit, the random vector set will have to include the vector shown in this figure. No other test vector will detect this specific fault. Thus, if the cascade of AND–OR is long enough, it will contain random-vector resistant faults. Because of these faults, we have to determine the fault coverage measure in pseudorandom testing in order to guarantee that the test quality will not be inferior to that of a deterministic test generation method.

The problem is reduced to selecting input test vectors that give adequate fault coverage. With pseudorandom testing, this involves selecting a number *L* of pseudorandom vectors and estimating the modeled-fault coverage that these vectors will yield. Mathematically speaking, *L* is selected so that $(100 - e)\%$ of the population's single-stuck-at faults would be detected with probability of $(1 - D)$, where *e* is the percentage of faults not detected and *D* is the probability for not detecting faults. Typically, it is preferred to have the value of *e* less than 2 and *D* less than 0.001.

The equality of any test is measured by the probability that the circuit will be judged fault-free when, in fact, it is faulty. This probability, frequently called the "escape probability," is a function of the following three factors.

1. *Detection probability profile of the circuit.* The detection probability of a stuck-at fault is defined as the probability that a randomly chosen test vector will detect the fault via the output data stream. Each circuit has its own detection probability profile that may contain certain random-vector resistant faults, such as occurs with redundant circuitry, self-masking topologies, high fan-in gates, or deeply leveled logic with some rarely sensitized paths.

2. *Test length.* For nonexhaustive testing, the test length L is closely related to the detection probability of the circuit and in particular to the hard-to-detect faults.

3. *Error masking.* If the compressed erroneous bit stream caused by a fault has the same signature as the correct one, then error masking occurs. Error masking should not be of primary concern because it is insignificant[10] when the data compressor uses multiple-input shift registers with sufficient width.

One result of nonexhaustive testing is that a large number of faults are detectable within the first few vectors, while subsequent vectors yield less and less additional coverage. Test quality can be improved by thoughtful partitioning of the circuit and through the use of special registers to be addressed in subsequent sections. But the determination of test quality is a difficult and necessary task. Traditionally, test quality can be evaluated by fault simulating the test vectors against the software model of the CUT. Unfortunately, this is a very time-consuming and costly process, especially if the number of vectors is very large. Recently, alternatives have been developed for fault simulation,[13] and the STAFAN method[14] in particular is efficient in estimating the detection probabilities for each stuck-at fault. It uses a modified logic simulator to simulate a sample of test vectors and to gather statistics from the results to make estimates of the fault coverage achieved. The computational complexity of STAFAN grows linearly with the number of circuit nodes and is implemented in the following manner.

- Special counters are added to the input and output of each gate of the circuit during logic simulation, including "zero" counters and "one" counters.
- As a sample of test vectors is simulated, logic values of the nodes are used to determine which counters get incremented.
- Detection probabilities are estimated for each stuck-at fault from the counter values. At this point, the percentage of fault coverage can be estimated; it equals the percentage of modeled faults with nonzero detection probability.
- The total number L of random vectors needed to detect $(100 - e)\%$ of the single, detectable stuck-at faults with probability $(1 - D)$ is calculated from the detection probabilities. Calculations generally yield usable values of L, usually substantially less than the total number of all possible input vectors.

5.2.3 Design Guidelines

In using an LFSR to generate pseudorandom tests, we recommend the following design guidelines.

- Always use primitive polynomials for random vector generation. As we discussed previously, test length used in nonexhaustive testing is closely related to the detection probability profile of the CUT and in particular to the hard-to-detect faults. This is why we emphasize the avoidance of nonprimitive polynomials in designing the LFSR so that maximal length of test vectors can be achieved.

- If test length L calculated is less than maximal length, use a maximal-length generator anyway and terminate the test by comparing the generator output to a predetermined vector.
- Use a logic simulator to verify generator outputs, identify the last desired test vector, and confirm the CUT signature.
- Use a reasonably complicated primitive polynomial for test data compression and test generation when both functions are performed by the same LFSR so that hardware overhead can be minimized. The additional chip area required to implement BIST should not exceed an objective of 5%.

5.3 Test Data Compression in BIST

For test verification to occur on-chip, the response data must be reduced to a manageable size. A number of alternative techniques are available as we discussed before. The most popular and error-sensitive technique is signature analysis.

In signature analysis, data compression through an LFSR reduces the amount of test response data to be handled from the entire data sequence into a single code word. This code word, called the signature or remainder, is the last remaining state of the compressing LFSR. Two events must occur for error detection to be made.

- The fault must be made visible at the output of the CUT.
- The signature must differ from that which would be produced from a data stream with no errors.

The first event depends on the generated test vectors and is independent of test data compression. The second event is best understood with further explanation of data compression and its polynomial division equivalence below.

5.3.1 Serial Input Data Compression

When a data stream enters an LFSR serially, it can be considered as the descending order coefficients of a "data-stream" polynomial. This phenomenon is called the polynomial division equivalence[3,5] of data compression and is illustrated by an example in Figure 5.13. Note that the characteristic polynomial of the LFSR shown below is defined using the notations in Figure 5.5.

$$X^5 + X^3 + X + 1$$

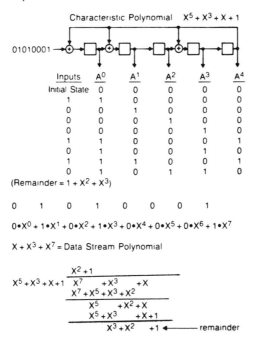

Figure 5.13   Polynomial division equivalence of data compression.

Also note that the input data stream of "01010001" corresponds to the data-stream polynomial below.

$$X^7 + X^3 + X$$

The remainder of the division of the data-stream polynomial by the characteristic polynomial is given below. This remainder corresponds to the final state in the memory devices in the LFSR.

$$X^3 + X^2 + 1$$

A faulty data stream will yield the same signature as an error-free data stream if and only if the faulty data-stream polynomial and the error-free data-stream polynomial have the same remainder when divided by the characteristic polynomial. The likelihood of error detection from a faulty data stream is heavily dependent on the compressor bit length r. For long input sequences, if we assume that all possible error patterns are equally likely, the probability of error detection [15] is $(1 - 1/2^r)$.

The assumption of uniformly distributed errors is not entirely accurate, but it does yield a figure of merit for comparing alternative compression techniques. This raises the interesting question of which polynomials are best suited for data

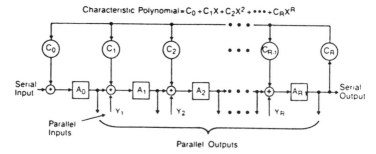

Figure 5.14 Multiple-input shift register IE-type.

compression. An interesting result is that any characteristic polynomial with two or more nonzero coefficients will detect all single-bit errors. The detection in itself would imply that the simplest choice, such as $(X^r + 1)$, may be an acceptable answer to our question. However, there are other considerations involved in choosing a suitable characteristic polynomial in the design of an LFSR for use in data compression.

Considerations must be given to dependent,[15] nonuniformly distributed errors such as burst errors, regular-interval errors, intermittent faults, and pattern-dependent errors. A more complicated feedback structure in the LFSR can more effectively prevent the masking of errors. The current trend is to use a reasonably complicated primitive polynomial for data compression.

5.3.2 Parallel Input Data Compression

Multiple-input shift registers (MISRs) are shown in Figure 5.14 for IE type and in Figure 5.15 for EE type. They are used in parallel input data compression because of their advantages in reducing the chip area and signature comparison times that would be required if separate compressors were used at each output of the CUT. The multiple inputs are the parallel inputs $Y_1, Y_2, \ldots, Y_r$. There is

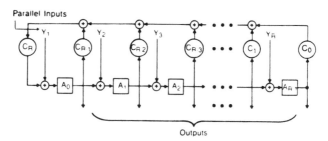

Figure 5.15 Multiple-input shift register EE-type.

an equivalence between the single-input and the parallel-input configurations. The inputs for an MISR when applied to a serial LFSR will give an identical output.

For external exclusive-OR MISRs, the equivalence with serial LFSRs is complicated, in that the equivalent data stream depends on the specific feedback connections and the current state of the shift registers. Nevertheless, as in the single-input case, it can be shown that the probability of detecting uniformly distributed errors in the output data stream[14] is approximately $(1 - 1/2^r)$.

5.3.3 Design Guidelines

In using LFSR for test data compression, we recommend the following design guidelines to reduce error-masking probability.

* For narrow output width r, the LFSR can be lengthened to m bits until a masking probability of $(1/2^m)$ is sufficiently small as illustrated in Figure 5.16.
* The test can be repeated using different feedback connections in the r-bit LFSR or repeated with the LFSR lengthened to $(r + 1)$ bits wide.
* Testing can be done by compressing the same test data into two different LFSRs concurrently.
* The test can be repeated using the same data compressor with a different vector set or a different vector sequence.
* Signatures can be computed and compared periodically during the test to capture errors that may be masked later in the test.
* The serial output of the MISR can itself be compressed into an LFSR to increase the probability of capturing errors.

Not all of the guidelines listed above are equally attractive. For example, to expand narrow output channels, the first guideline is definitely appropriate. But the second guideline may present routing and physical layout problems. Each of

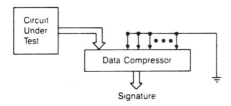

Figure 5.16 Expanding a small output channel to reduce masking probability.

the guidelines requires additional efforts by the designer. In the end, the benefit will outweigh the extra time and effort spent in the design phase.

5.4   The BILBO Registers

In this section, we will present a special LFSR configuration which is very popular in BIST. It combines the regular features of built-in self test and the scanpath design methodology. It is a general purpose LFSR called built-in logic-block-observation (BILBO)[16] that was introduced in the late 1970s and is now widely used in BIST applications. BILBO can also be used as latches and shift registers during the normal mode operation. An 8-bit multiple-input EE type configuration is illustrated in Figure 5.17.[3] The characteristic polynomial for the LFSR is given below.

$$X^8 + X^5 + X^3 + X + 1$$

In this figure, the Z_i's are parallel input lines and the Q_i's are register output lines. Both parallel and serial input/output are possible. There are four operating modes under the selection by the input values of $B_1 B_2$ for the BILBO register. Depending on the circumstances, only some of the modes may be useful. In those cases, simplification of the basic structure would result in some hardware savings. The four modes are defined below.

Mode 1, $B_1 B_2 = 11$, is the basic operating mode for normal register function and results in the configuration shown in Figure 5.18. Note that the registers are decoupled from each other with input Z_i and output Q_i. These registers can be used in any manner required by the design of the circuit.

Mode 2, $B_1 B_2 = 00$, is the serial LFSR mode as shown in Figure 5.19. The

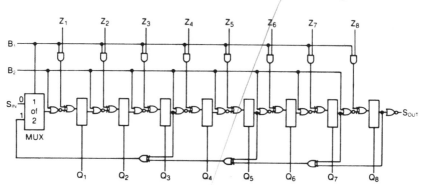

Figure 5.17   A BILBO register.

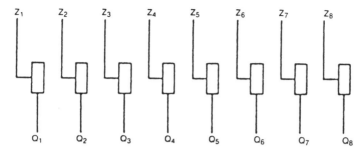

Figure 5.18 Mode 1: $B_1B_2 = 11$.

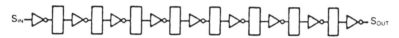

Figure 5.19 Mode 2: $B_1B_2 = 00$.

LFSR can itself be tested by shifting in and out a known sequence of data stream. It is also used to serially clock out a final signature after testing. If no other reset mode is available, mode 2 can be used for initializing a sequential CUT by shifting in the required initial state of the registers.

Mode 3, $B_1B_2 = 10$, is the multiple-input signature analysis mode as shown in Figure 5.20. Besides the function of a parallel signature analyzer, it can also be used for generating pseudorandom test sequences. This is achieved by keeping the inputs Z_i fixed on any desired logical values.

Mode 4, $B_1B_2 = 01$, is a register reset mode. It forces all registers to all zeros.

Using the BILBO circuitry, the general BIST approach in Figure 5.1 can now be redrawn as Figure 5.21. To utilize the BILBO concept further, all of the memory devices in a sequential circuit can be included in BILBO registers as shown in Figure 5.22. This leaves only the combinational logic for pseudo-random testing. This greatly simplifies the task of obtaining the high fault cov-

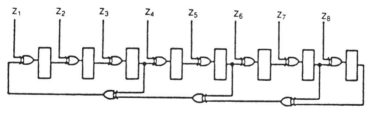

Characteristic Polynomial $X^8 + X^5 + X^3 + X + 1$

Figure 5.20 Mode 3: $B_1B_2 = 10$.

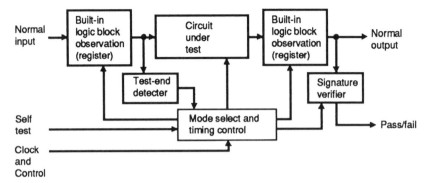

Figure 5.21 A general BIST configuration incorporating BILBO registers.

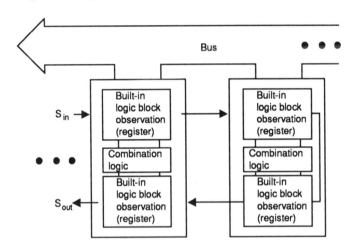

Figure 5.22 A BILBO configuration including all memory elements.

erage required for BIST. This configuration also allows output of several signatures in a partitioned circuit by switching from mode 3 to mode 2. Circuits partitioned for BIST will be discussed in the next section.

5.5 Subcircuit Partitioning and Diagnostics

A useful first step for designing built-in self test is to partition the circuit into subcircuits in order to increase fault coverage in a nonexhaustive test. This is done by reducing the size of the individual circuits to be tested. Guidelines for partitioning fall into three categories: design objectives, general requirements, and specific techniques.

5.5.1 Partitioning Objectives

Partitioning decisions should be made with considerations given to the following objectives.

- Minimize BIST overhead
- Minimize the performance degradation, especially for speed in the normal operation
- Incorporate the memory elements in BILBO registers
- Maximize fault coverage and use exhaustive testing if necessary
- Maximize the speed of BIST process
- Allow partitioning to achieve a useful level of fault isolation

Note that sometimes these objectives may be in conflict with each other for a particular design. Then some compromises would have to made that best suit the given circumstances.

5.5.2 Partitioning Requirements

Since in BIST each CUT must have a known fault-free signature for any given input sequence, the following general requirements are specified to ensure this.

- No logic indeterminacy allowed (i.e., no "don't cares" are allowed in any partitioned subcircuits). Indeterminate values can leave an LFSR in an un-predictable state.
- Memory elements of the CUT must be initialized before testing and the partitioning should not leave any of them noninitializable. Optimally, all memory elements will be included in BILBO registers, leaving only com-binational logic for pseudorandom testing.
- Race and hazard conditions must be avoided. Pseudorandom test vectors may produce test conditions that are never intended to occur during normal operation. Also, simultaneous switching limits may be unexpectedly ex-ceeded in a testing environment that forces storage elements to toggle.

Note that any one of the requirements when violated in a given BIST design can lead to invalid test results. As such, they should not be treated as design objectives but instead as design rules to be followed faithfully.

5.5.3 Specific Partitioning Techniques

There exist many techniques for circuit partitioning, but the following are appli-cable to a BIST circuit in particular.

- Partition the VLSI chip so that the logic within a partition comprises a func-tional entity (i.e., partition the chip functionally).
- Choose partitions so that the subcircuits can be tested independently and, if desired, in parallel with each other.

- Make separate partitions for higher failure rate logic, critical logic, or logic having a higher than average usage rate. Also make separate partitions for storage elements whose writing-to-reading times are relatively long.
- Create partitions to provide any required level of thoroughness for individual fault detection to augment BIST's limited fault location capabilities.
- Ensure that each partition has no more inputs than can be processed during the time allocated for that partition.
- Consider merging a partition with relatively few inputs with another partition.
- Extend the compressor width if a partition has few outputs that may result in unacceptably high error-escaping probability.

5.5.4 BIST Diagnostics

BIST practically has no inherent diagnostics capabilities. They are basically go/no–go testing techniques that allow for design speed production testing. Two approaches[4] to enhance the diagnostic capability of BIST are discussed below. Both approaches identify fault location to within the cone of logic feeding a particular compressor output line.

The first approach, shown in Figure 5.23, is to rerun the test for each individually selected CUT output line if an erroneous signature is first received. Based on the information regarding which output line is producing the error bit, the fault can be isolated to the circuit logic responsible for the fault.

The second approach, shown in Figure 5.24, involves inspection of the portion of the output matrix that contains the first detectable error stored temporarily in the RAM. This requires periodic inspection of the current signature during the test run and determination of the failing vectors located in the RAM. Again based on the failing output line information as well as the information regarding the vectors that cause the failure, fault isolation and location is possible.

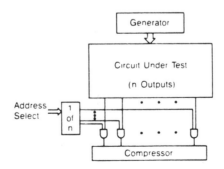

Figure 5.23 Fault isolation by CUT's output lines.

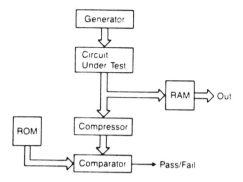

Figure 5.24 Temporary local storage of failure-related information.

5.6   BIST Feature Summary

Advantages and disadvantages of built-in self test are summarized in this section. Where appropriate, they are contrasted with the scan-path design and testing methodology.

- With BIST, vectors are generated automatically inside the chip nearly at design speed. No other testing techniques, including scan-path, have this property.
- Neither test nor response vectors need to be stored external to the chip. Consequently, a chip can be tested in the field[7] without relying on expensive automatic test equipment.
- BIST gives no indication of the type of failure in the chip and therefore is not useful in the development process of the design. In contrast, scan-path is a deterministic fault-locating technique.
- Both scan-path and BIST make good utilization of silicon area in that they contain hardware used for normal operation and for testing.
- As a rule of thumb, a reasonable design for BIST should not consume more than 5% of the silicon area. This makes BIST an increasingly attractive testing technique at higher and higher levels of integration.
- By partitioning of a CUT, testing efforts with BIST can be made practically independent of chip density. Furthermore, testing time can be reduced by running several tests concurrently or by running several tests before signature examination.
- BIST is most useful for designs with regular arrays of latches and registers[17] and for pipeline-type circuits. Scan-path is good for scattered, independent single flip-flop designs.
- The BIST techniques discussed are directly adaptable to higher levels of

assembly such as system-level verification test[18] and have been successfully employed in field testing of digital instrumentation.

Because of the advantages of BIST in cost, speed, and flexibility, it will be the required design-for-test method in the future.

5.7   Boundary Scan Overview

Boundary scan is a design-for-testability technique intended to facilitate testing primarily of printed circuit boards or higher level assemblies of digital circuits or systems. It was formulated[19] by the Joint Test Action Group (JTAG) and is standardized[20] as IEEE Standard 1149.1. In this section, we will introduce the basic concepts behind the boundary scan technique, its building blocks and architecture, and operational requirements.

5.7.1 Introduction

Testing of individual circuits and their interconnections on a printed circuit board (PCB) is accomplished through access to individual I/O pins using the bed-of-nail fixtures of an in-circuit tester. With the increasing number of circuits on the board and the extensive use of surface-mount technology, access to these I/O pins is getting more difficult. As a result, the in-circuit testing method is losing its effectiveness.

The boundary scan technique is proposed as a design-for-testability framework with a standardized architecture to allow easy access to individual circuits on a board for testing and also to allow access to testing features embedded in these circuits (e.g., BIST). Fundamental to its architecture is a boundary scan cell containing a shift-register latch for each I/O pin of a circuit to allow signals at the boundaries of the circuit to be completely controllable and observable. Each boundary scan cell is able to capture data either from an input pin or from the circuit logic, and it can load data either into the circuit logic or onto a circuit output pin.

The boundary scan cells in a circuit can be interconnected into shift registers along the boundary of the circuit and thus provide a path to serially shift data into and out of the shift register under the control of a clock and control signals. The shift register path of a circuit can be extended further to a complete shift-register chain on a board by serially connecting the shift registers in other circuits of the board. This forms a boundary scannable board design, as shown in Figure 5.25. This figure presents a simplistic view of the boundary scan structure and shows its equivalence to the scan-path structured DFT technique for an integrated circuit (IC).

Note that the heavy lines are used to indicate the shift-register chain formed,

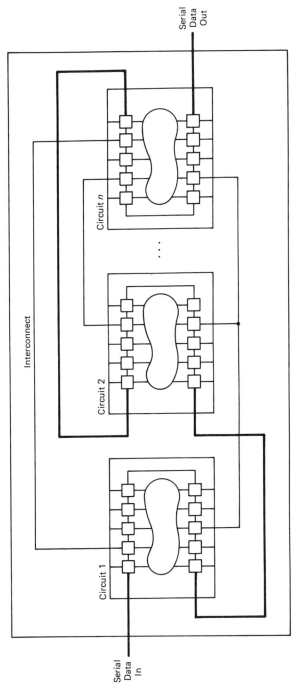

Figure 5.25 A boundary scan board.

and it uses a single scan-in pin and a single scan-out pin on the edge of the board. This provides a framework for board-level testing which allows serial access and control from outside of all the primary I/O pins of ICs on the board. As long as the design conforms with IEEE 1149.1, the ICs do not have to come from the same vendor and the boundary scan design objectives can still be achieved.

The testing objectives that can be achieved using the boundary scan technique are as follows.

- Testing of the interconnections between various ICs on a board. Test data can be shifted into all the boundary scan cells for an IC's output pins and then loaded in parallel through the interconnects to the boundary scan cells for all the input pins of the ICs that are interconnected. The test results can be verified by shifting out the data received from the boundary scan cells for the input pins through the serial data output port of the board.
- Testing of individual ICs on a board. Boundary scan structure allows the control of BIST operations through commands and control signals shifted into an IC. The boundary scan registers can be used to isolate the logic under test from stimuli received from surrounding ICs.
- Using boundary scan registers to perform other testing functions. These registers can be designed to perform the additional functions of generating pseudorandom test vectors and of signature analysis in a BIST application. They also permit a limited slow-speed static test of the system logic of an IC because test data can be shifted in and test results can be shifted out.

As we can see, the boundary scan concept and structure provides a sort of electronic in-circuit testing facility. In addition, it provides the necessary features for implementing BIST and other types of tests.

5.7.2 Basic Architecture

The overall boundary scan architecture[19,20] as shown in Figure 5.26 consists of a test access port (TAP) which includes a five-signal interface, a controller, an instruction register, and a variable number of test data registers. A test data register contains the boundary scan register, formed by serially linking registers in the boundary scan cells that are adjacent to each circuit's I/O pins. It also contains a bypass register and an optional device identification register. The boundary scan registers in circuits of a board are linked into one or more serial paths through the board to allow (1) the test of interconnects, (2) the applications of tests of each circuit in the path and also clusters of logic not in the path, and (3) access to testing features such as BIST. It also allows the monitoring of the system normal operation by taking snapshots of the states of circuits on the board.

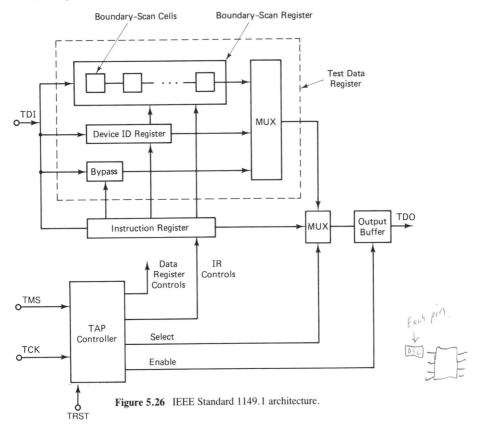

Figure 5.26 IEEE Standard 1149.1 architecture.

The elements of the boundary scan architecture are as follows.

- *Test access port:* A 16-state FSM which receives test clock signal (TCK) and test mode select (TMS) input from the IEEE Standard 1149 testability bus. The controller's primary function is to generate clock and control signals for the rest of the architecture. An optional reset may also be provided as an input signal for the controller.

- *Test clock signal:* One of the input signals. A separate clock input is provided so that independent and concurrent shifting of test data is possible. It ensures that test data can be moved from or to an IC without changing the state of the system logic.

- *Test mode select signal:* One of the input signals. The value of TMS determines state transitions for the FSM that has three major types of states that represent the basic actions required for testing: stimulus application, execution, and response capture. All state transitions of the TAP controller must occur based on the value of TMS at the time of the rising edge of TCK.

- *Test data input (TDI) signal:* One of the input signals. TDI provides serial movement of test data through the board. The values present at TDI clocked into either the instruction register or the test data register at the rising edge of the clock signal TCK.

- *Test reset (TRST) signal:* One of the input signals. It is an optional signal in the TAP controller design. If it is included, it provides an asynchronous reset to the FSM when enabled. However, TRST must not be used to initialize any system logic within the circuit.

- *Test data output (TDO) signal:* The output signal. Its primary function is to shift out test data in the selected register (either instruction register or test data register). The capability of the TDO pin switch between enable and disable is required to allow parallel, rather than serial, connection of test data paths when needed. To ensure race-free operation of the boundary scan techniques, the changes at the TDO signal must be delayed until the falling edge of the TCK signal.

- *Instruction register (IR):* A multibit register that receives an instruction from the test data input (TDI) and which selects the test to be performed, such as external, internal, or sample. It also selectively accesses a register in the data register such as the boundary scan, bypass, or device identification register.

- *Boundary scan register (BSR):* A multibit shift register. This is formed by interconnecting boundary scan cells serially with access to the circuit's I/O pins and internal logic. It allows testing of interconnects between circuits to detect faults associated with opens, shorts, and stuck-at. It also allows testing of a circuit's system logic and sampling of signals passing through the circuit's inputs and outputs.

- *Bypass register (BR):* A single-bit register. This provides a connection from TDI to TDO to allow test data to pass through to other circuits on the board with one unit of delay without affecting normal system operation of the particular circuit.

- *Device identification register (DIR):* An optional multibit test data register. It allows the manufacturer, part number, and variant of part to be shifted in and stored as a binary identification code.

Note that the BSR, BR, and DIR described above form the test data register (TDR). Within a TDR, circuitry may be shared with other TDRs as long as the rules in the IEEE Standard 1149.1 are not violated. Unless stated otherwise in the standard, circuitry in TDRs can also be used to perform normal system logic when testing is performed. In addition, other optional test data registers intended for accessing privately held testing features can be included in the architecture. They may be made publicly accessible at the discretion of the IC manufacturer.

5.7.3 Operating Modes

The boundary scan register is the principal building block of the boundary scan technique. It is a parallel-in, parallel-out shift register and can be viewed as a four-terminal device as shown in Figure 5.27. This view of a boundary scan register cell is to facilitate the description of the flow of data through these cells when they are configured in different modes of operation as determined by the instruction shifted into the instruction register.

There are the following major operating modes of the boundary scan technique: normal, scan, external test, internal test, sample test, and built-in self test.[21] The mode of operation of each test data register in each TAP controller state is entirely defined by the current instruction in the instruction register. We will describe the various modes of operation in terms of the configuration of the 4-bit input boundary scan register in Figure 5.28.

In the normal mode of operation, the circuit performs its designed functions. The input boundary scan register must appear transparent to the input operational data. The flow of data between a pair of input and output boundary scan register cells is shown by the heavy line in Figure 5.29.

In the scan mode of operation, the boundary scan cells must form a shift-register chain by interconnecting them serially. The flow of data in this configuration is shown by the heavy line in Figure 5.30.

In the external test mode of operation, the circuitry external to the circuit, the interconnects between ICs typically are tested. Test data is applied to the output boundary scan register in one chip, and the input boundary scan registers in the interconnecting chips latch the data flowing between them. This data can then be shifted out for verification. The flow of data in this configuration is shown by the heavy lines in Figure 5.31.

In the sample mode of operation, the circuit performs its normal mode of operation and meanwhile its signal flow is concurrently being monitored. Data

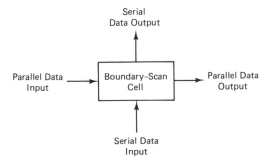

Figure 5.27 A view of a boundary scan register cell.

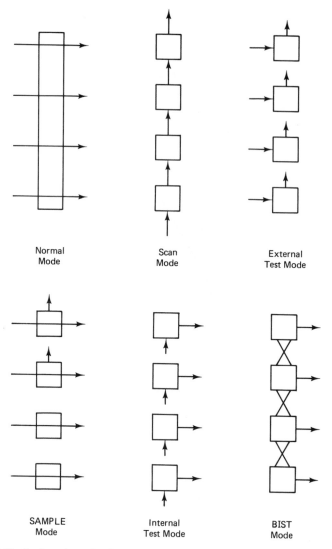

Normal
Mode

Scan
Mode

External
Test Mode

SAMPLE
Mode

Internal
Test Mode

BIST
Mode

Figure 5.28 Configurations of an input boundary scan register for various modes of operation.

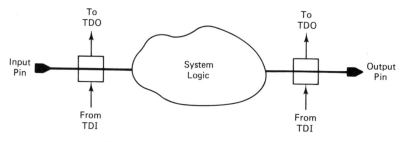

To
TDO

To
TDO

Input
Pin

System
Logic

Output
Pin

From
TDI

From
TDI

Figure 5.29 Data flow in the normal mode.

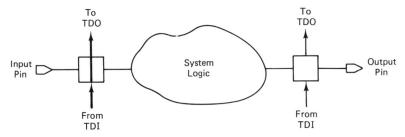

Figure 5.30 Data flow in the scan mode.

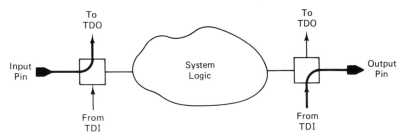

Figure 5.31 Data flow in the external test mode.

is latched by the input and output boundary scan registers. A snapshot can be taken at any instant for diagnosis and maintenance. This mode thus allows fault localization to system logic between the pair of input and output boundary scan register cells. It also allows the testing of interactions between ICs on the board. The flow of data in this configuration is shown in Figure 5.32 by heavy lines.

In the internal test mode of operation, the system logic internal to the circuit can be tested. Test data is shifted into the input boundary scan register and applied to CUT. The response data are then latched by the corresponding output boundary scan register to be shifted out for verification. During internal testing of a given circuit logic, the logic level at the output pin can be defined through its associated boundary scan register to ensure that proper logic values are pro-

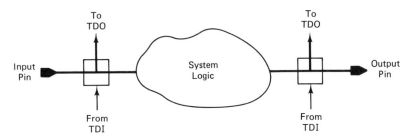

Figure 5.32 Data flow in the sample mode.

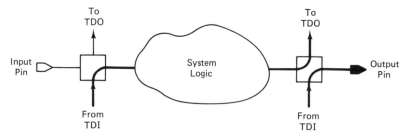

Figure 5.33 Data flow in the internal test mode.

vided to the connecting circuits. The flow of data in this configuration is shown in Figure 5.33 by heavy lines.

In the built-in self test mode of operation, the internal self-test features are executed. While this mode is selected, the boundary scan register may also be configured as a pseudorandom test vector generator or as a test-results compressor for signature analysis. The functions of this register are similar to the BILBO register[16] addressed in Section 5.4.

During self testing of a given circuit logic, the logic level at the output pin can be defined through its associated boundary scan register to ensure that proper logic values are provided to the connecting circuits. Note also that during internal testing, the boundary scan register cells at the input pins may be allowed to provide user-specified logic levels to be established at the system logic inputs. This can shield the system logic from any unwanted signals coming from the interconnects that may interfere with the internal test that is in progress. The flow of data in this configuration is shown in Figure 5.34 by heavy lines.

5.8   Boundary Scan Applications

With the background information of the JTAG/IEEE Standard 1149.1 boundary scan technique in the last section, we will present several cases of applications. We will first discuss the use of a boundary scan chain to test conventional logic

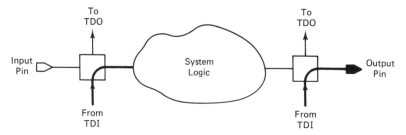

Figure 5.34 Data flow in the built-in self test mode.

and memory clusters on a board. We will then discuss applications of the boundary scan technique to the testing of interconnects, as well as to the implementation of BIST. Design tradeoffs of some of the related subsystems will also be discussed.

5.8.1 Testing of Conventional Logic

Although many ICs will be designed with boundary scan features to facilitate testing in the next few years, it is not realistic to expect all devices on a board to be boundary scannable. This is especially true for small commercial logic devices where economical considerations may outweigh the need to incorporate design-for-test features. Instead of seeing full boards of boundary scan devices in the near future, a more likely scenario will be a prevalence of boards with both boundary scannable ICs and conventional logic and memory devices with no scan capability at all. Testing of these mixed-mode boards presents a new challenge for the testing engineers. A promising new approach was recently proposed that uses the I/O pins for boundary scan ICs as virtual ATE channels[22] to test surrounding nonscannable devices on the same board.

With reference to Figure 5.35, test vectors to test these nonscannable devices are shifted through the scan path to the output boundary scan register cells of

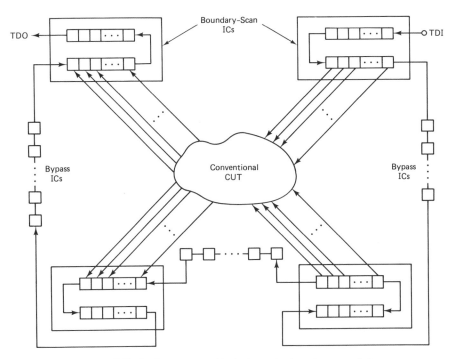

Figure 5.35 Testing nonscannable devices using boundary scan chain.

scannable devices that surround the conventional CUT. The outputs then drive the data into the input pins of the CUT. The test responses are then captured by the input boundary scan register cells of other scannable ICs to be shifted out through the boundary scan chain on the board for analysis and verification.

Although this seems to be a straightforward approach, there are testability requirements to be considered in order to make this approach a success. One particular consideration is the limitation of the speed to apply test vectors since all test data have to be shifted in serially into the output boundary scan cells. This may pose problems for dynamic logic which needs frequent freshness. Test points need to be added so that dynamic ATE channels can be used to provide the refreshing signals. A second consideration is to avoid large sequential depth to make test generation for the conventional CUT easier, since either ATVG or manual test generation are used to produce the necessary vectors to test these circuits.

To test these clusters of conventional logic on a board, two scanning operations are needed, one for vectors and one for responses. However, these two operations can be carried out in parallel. When the next test vector is being shifted into the scan chain, the responses for the current vector can be shifted out concurrently.

Another problem associated with this approach is the huge amount of data required to test clusters of conventional logic and also memory devices. Using special hardware and data compression techniques, the storage requirements for the test data can be made small and the test time minimized to the delay along the boundary scan path.

5.8.2 Testing of Interconnects

One of the major advantages of incorporating the boundary scan design method on a board is in the testing of interconnects, because techniques can be developed to facilitate fault detection[23,24,25] and diagnosis without the need of the bed-of-nail features as in testing the conventional logic. Stuck-at and bridging faults, the two most common types of faults associated with interconnections between ICs on a board, are both amenable to test vectors generated taking advantage of the structure provided by the scan architecture.

Typical interconnects on a board are shown in Figure 5.36, where I/O boundary scan cells are connected with equal-potential electrical wires called nets. The simplest interconnect is a pair of scan cells connected by a net, where the output register is referred to as the driver and the input register the receiver. Figure 5.36a shows several parallel nets and each of the nets can be considered as the net-under-test (NUT). Fanout results when an output boundary scan cell is connected to more than one input scan cell (Figure 5.36b). But this interconnect structure is considered as having only one net.

Figure 5.37 shows a net driven by more than one driver and is called a wired-

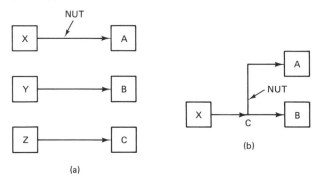

Note: X, Y, Z are output boundary-scan
register cells.
A, B, C are input boundary-scan
register cells.

Figure 5.36 Different interconnect structures: (a) Parallel interconnect net;
(b) fanout interconnect net.

AND/OR net. When the output buffer can be controlled by a disable input, then
this becomes a tristate interconnect net as shown in Figure 5.38. Each of these
two interconnect structures is considered as having one net. Depending on the
driving characteristics of the output cell, the wired type of net can be either a
wired-AND (where the resultant logic value is an AND of logic values of individual nets) or a wired-OR (where the resultant logic value is an OR of logic
values of individual nets). For a wired-AND net, a 0 dominates; for a wired-OR
net, a 1 dominates. More complex interconnects made up of the types of interconnects we have discussed so far can also be found on boards.

There are two types of interconnect faults of interests, bridging faults and

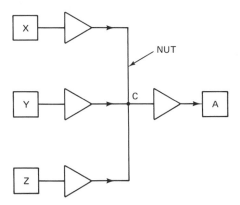

Figure 5.37 A wired-AND/OR interconnect net.

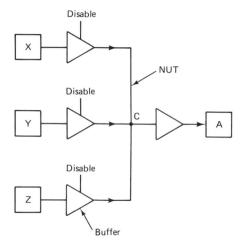

Figure 5.38 A tristate interconnect net.

stuck-at faults. Bridging faults are multinet faults which basically cause a short between two or more nets. Stuck-at-0, stuck-at-1, and open faults can occur on single nets, wired nets, or tristate nets. Note that stuck-at faults affect the wired or tristate nets as a whole while open faults may affect only part of the nets, as we discussed in Chapter 1. But an open fault on a single net is equivalent to a stuck-at-0 or stuck-at-1 fault at the receiving buffer, depending on the technology. We will address the testing of the stuck-at faults on the wired-AND/OR nets first and then the testing of bridging faults on more than one net.

5.8.3 Testing of Stuck-At Faults

P.T. Wagner found an algorithm for testing stuck-at faults on wired nets and bridging faults between multinets. The discussion of testing procedures of these faults in this and the next section are based on his work.[23]

The testing of stuck-at faults on a wired-AND net is similar to that of testing an AND gate as discussed in Chapter 1. If there are n drivers, then a total of $(n + 1)$ vectors are needed to detect all the stuck-at faults. This is done by walking a 0 across the output buffers while the remaining $(n - 1)$ output buffers are held at 1 and the extra vector consisting all 1s is for testing the stuck-at-0 faults. The dual procedures apply to the testing of wired-OR nets. To reduce test time, multiple wired nets can be tested in parallel and a total of $(n + 1)$ are adequate to test all wired AND/OR nets. The step-by-step testing procedures are presented below for the stuck-at and open faults on wired nets.

The procedure for testing a wired-AND interconnect net for stuck-at-1 faults is as follows.

1. Shift a logic value 0 into a driver (e.g., X in Figure 5.37) that drives the NUT.
2. Shift a logic value 1 into all other drivers (e.g., Y and Z in Figure 5.37).
3. The vector is clocked into the receivers (i.e., A in Figure 5.37).
4. Shift out the test data from all the receivers on the NUT and examine for a logic value 0.
5. The test is passed if no logic 1 is received in the test data.
6. Repeat the above steps for each of the drivers on the NUT.

The procedure for testing a wired-AND interconnect net for stuck-at-0 faults is as follows.

1. Shift a logic value 1 into all drivers that are associated with the NUT.
2. The vector is clocked into the receivers.
3. Shift out the test data from all the receivers on the NUT and examine for a logic value 1.
4. The test is passed if no logic 0 is received in the test data.

The procedure for testing a wired-OR interconnect net for stuck-at-0 faults is as follows.

1. Shift a logic value 1 into a driver that drives the NUT.
2. Shift a logic value 0 into all other drivers.
3. The vector is clocked into the receivers.
4. Shift out the test data from all the receivers on the NUT and examine for a logic value 1.
5. The test is passed if no logic 0 is received in the test data.
6. Repeat the above steps for each of the drivers on the NUT.

The procedure for testing a wired-OR interconnect net for stuck-at-1 faults is as follows.

1. Shift a logic value 0 into all drivers that are associated with the NUT.
2. The vector is clocked into the receivers.
3. Shift out the test data from all the receivers on the NUT and examine for a logic value 0.
4. The test is passed if no logic 1 is received in the test data.

In a tristate interconnection only one driver can be active at a given time. Therefore, the testing of stuck-at faults on a tristate interconnect net is similar to that of testing a wired-AND net for stuck-at-1 or a wired-OR for stuck-at-0 faults, except that each driver must be tested while the rest of the drivers are disabled.

The procedure for testing a tristate interconnect net for stuck-at-0 faults is as follows.

1. Shift a logic value 1 into an enabled driver that drives the NUT.
2. Shift a logic value 0 into all other disabled drivers.
3. The vector is clocked into the receivers.
4. Shift out the test data from all the receivers on the NUT and examine for a logic value 1.
5. The test is passed if no logic 0 is received in the test data.
6. Repeat the above steps for each of the drivers on the NUT.

The procedure for testing a tristate interconnect net for stuck-at-1 faults is as follows.

1. Shift a logic value 0 into an enabled driver that drives the NUT.
2. Shift a logic value 1 into all other disabled drivers.
3. The vector is clocked into the receivers.
4. Shift out the test data from all the receivers on the NUT and examine for a logic value 0.
5. The test is passed if no logic 1 is received in the test data.
6. Repeat the above steps for each of the drivers on the NUT.

The diagnosis of a stuck-at fault is rather easy because the affected receiving buffer contains a constant value instead of the value of the driver. However, care must be taken to achieve complete diagnosis between stuck-at and bridging faults. For better fault resolution, additional test vectors are more frequently needed than proposed above and this will be discussed in the next section.

5.8.4 Testing of Bridging Faults

In this section we will discuss the testing of bridging faults between two nets that are electrically shorted. This can occur between any two types of nets including multidriver and single-driver nets as shown in Figures 5.36 through 5.38. We will discuss testing for these faults in terms of their detection, isolation, and diagnosis.

The procedure for testing a bridging fault between two interconnect nets is as follows.

1. Shift a logic value 0 into all the drivers of one of the two NUTs.
2. Shift a logic value 1 into all the drivers of the other NUT.
3. Enable all drivers on the two NUTs.
4. The vector is clocked into the receivers.
5. Shift out test data from all the receivers on each NUT and examine for a logic value 0 or 1, depending on which NUT the receivers are associated with.
6. A short exists between the two nets if at least one receiver contains data that is different from the value transmitted by its driver(s).
7. Repeat the steps for each pair of interconnect nets.

The vector set thus generated contains[26] $\log_2(k + 2)$ test vectors where k is the number of interconnect nets to be tested for bridging faults on a board. As an example, we assume that there are 4 interconnect nets to be tested for bridging faults. The total number of test vectors needed is $\log_2(4 + 2)$ or 3. The set of three parallel test vectors (PTV) needed to detect the bridging faults between each pair of interconnect nets is listed in Table 5.1. Each of the three vectors is applied to all the interconnect nets in parallel.

These vectors are generated by assigning a number to each net and then assigning the 3-bit binary code for the number horizontally in the table. For instance, net 1 is assigned (001) and net 2 is assigned (010), etc.

Note that between each pair of nets at least one of the parallel test vectors has a logic value different for those two nets. For instance, PTV_3 has a 0 for net 2 and a 1 for net 3. As a result, PTV_3 detects the short between net 2 and net 3. The test vector set generated using the previous procedure can detect the existence of all the bridging faults among interconnect nets.

However, sometimes the same faulty responses can be due to either a stuck-at fault or a bridging fault. Care must be taken to distinguish between them. For example, if net 3 and net 4 are shorted, then the responses to the test vector set in Table 5.1 will be (111) for both nets, which is the same as if both nets are stuck-at-1. Distinction can be made[25] if an all-zero vector is added for wired-OR types of shorts or an all-one vector for wired-AND types of shorts. In Table 5.1, we can add one additional PTV containing all 0s. Note that a stuck-at-1 fault will drive the receiver to a 1 while a short will drive it to a 0. This will distinguish between the two types of faults.

A scan-path structure independent test was also proposed[24] to simplify test generation for testing interconnects. The total number of test vectors proposed is $\log_2(L + 2)$, where L is the number of boundary scan register cells in the scan path. The advantage of this scheme is that the test sets are order-independent and therefore suitable for pseudorandom testing in a BIST implementation. But it works only if the CUT does not contain tristate-interconnect nets.[25] If such nets exist, then the pseudorandom tests generated may cause conflicts between the tristate–disable input and the outputs they control, because both lie on the same

Table 5.1

Test Vectors to Detect Bridging Faults

Net number	PTV_1	PTV_2	PTV_3
1	0	0	1
2	0	1	0
3	0	1	1
4	1	0	0

scan path. The relationship between the tristate control signal and the output must be deterministic but not random.

5.9   Boundary Scan Implementation Considerations

Industry today is rapidly accepting IEEE Standard P1149.1 for implementing board-level or higher level design-for-testability features. In this section we discuss some of the design considerations and implementation experience in recent applications [27,28] of this standard.

5.9.1 Boundary Scan Controller

Because the boundary scan path provides access to the primary I/O pins, the board-level testing process basically consists of serially shifting testing data in and out of the CUT as well applying the tests by executing the necessary clock operations. This process eliminates the need not only for a bed-of-nail testing fixture for probing into the primary I/O but also for expensive automatic test equipment. An inexpensive host test computer together with the TAP controller and a well-designed interface between the host and the controller can perform the necessary functions of a regular ATE. These pieces form the boundary scan environment that performs the major functions of applying protocol commands to control the operation of the boundary scan ring, applying test data, and monitoring test results.

S. Vining [27] outlined a set of design tradeoffs of the control environment for a boundary scan implementation using a personal computer as the host test computer. The basic control environment consists of a host test computer, a controller, and a bidirectional parallel interface bus between the host and the controller, as shown in Figure 5.39. Note that the parallel interface is much faster than the serial interface. This implementation requires the controller to accept parallel instructions and test data from the host testing computer and then convert them into the serial form specified by IEEE P1149.1. Since scanning is a very time-consuming activity, further speed improvement can be accomplished by implementing bit manipulation and shift operation in the hardware of a dedicated controller. The controller can be implemented on a card that can be inserted in the host or placed on the board under test. In this environment, the primary functions of the host testing computer are to apply test instructions and test data, as well as to monitor test results. The controller itself acts as a smooth interface between the testing computer and the scan ring on the board under test by generating scan protocols and formatting serial data.

Design considerations and tradeoffs in the implementations of the boundary scan control environment are addressed here. The discussion will be centered in two major areas: host computer interface and scan operation throughput.

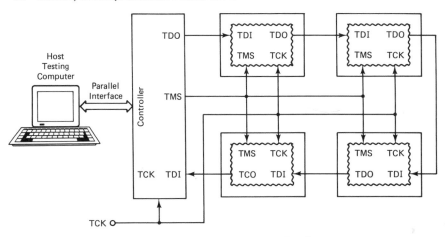

Figure 5.39   A basic boundary scan control environment.

A major design consideration in the interface between the host testing computer and the controller is the required synchronization between the test clock (TCK) and the host processor's write and read operation to ensure proper test execution. In particular, parallel bits of a command from the host processor should not change during clock transitions. Write synchronization should take place between write pulses, not during write pulses, to avoid lengthening the write pulse duration. As a result, the data from the host are latched by the controller hardware within a small number of test clocks. The processor read operation should also synchronize with data and status updates in the controller so that they will not change during a read operation. When the processor desires to read the new status and data, it issues an update command to cause the controller to load such data into the latches to be read. The data in the latches remain unchanged between update commands so that no data will be missed.

The scan throughput in a boundary scan implementation is mainly determined by the scanning rate (i.e., how fast serial test data can be shifted in and test results can be shifted out). Hardware implementation of scan-related features in the controller can improve scan efficiency, but at the cost of extra area and other overhead in the controller design. As in a general scan path implementation, the layout of an IC also impacts the serial path.[29] The boundary scan register cells can be connected into a serial chain not in the exact logic sequence from between the most-significant bit and the least-significant bit. Their order of connection can be varied for the sake of a shorter route or other layout and routing considerations. Note that the other nets not in the scan path should remain in the same order as required in the normal mode of operation.

Other implementation considerations include the breadth of the scan command set, which also impacts the efficiency of scan and test execution. A command set of the host testing computer broad enough to include commands to write and read data, to apply a single test vector or step through a test, and to reset the controller is needed for proper and fast operation of a boundary scan testing environment.

5.9.2 BIST Implementation

As shown in Figure 5.28, the boundary scan registers can be configured into the built-in self test mode. While in this mode, the boundary scan registers are configured as a pseudorandom test vector generator and a test-results compressor for signature analysis. One particular implementation involves the use of BIST within the boundary scan architecture to test a memory board.[30]

Figure 5.40 shows the block diagram of a memory board with RAM or EEPROM together with the added boundary scan circuitry organized into four subsystems: the global bus interface, BIST controller, memory decoding logic,

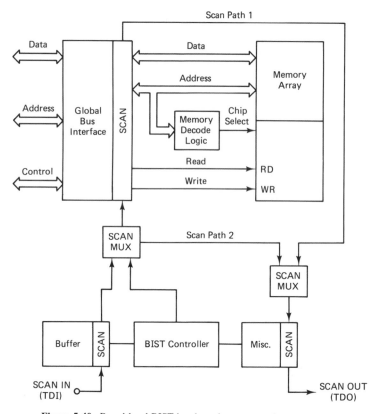

Figure 5.40 Board-level BIST in a boundary scan environment.

and the memory array. Boundary scan register cells are designed into data, address, and control signals so that a scan path (i.e., blocks labeled SCAN in the diagram) can be formed to access all the subfunctions on the board through the standard boundary scan interface.

Multiplexers in the scan path allow the control of the global bus interface either by the BIST controller or by the external TAP controller. The BIST controller itself is implemented as a finite state machine and stored in ROM. The self-test sequence is started after the BIST is enabled by the TAP controller, which also isolates the internal scan path from any external scan path so that the BIST controller can shift data into the scan registers in the global bus interface subsystem. The functional flowchart in Figure 5.41 shows how the BIST FSM controller handles the testing of all the memory arrays on the board.

During steps 2 through 5, the TAP controller polls the buffer periodically to see if the "testing complete" indicator has been set. If it detects this signal, then it de-asserts the BIST enable and SCAN MUX reconfigures all scan registers to the external path. As indicated in step 5, the signature contained in the parallel signature analysis register embedded in the data buffer in the global bus interface is scanned out for comparison with the expected value. If the two values do not agree, then a fault exists in the memory being tested and diagnostic testing must follow to localize the fault.

Valuable experience has been gained in using BIST with boundary scan. One particular lesson learned is that when BIST is inactive, it also can disable the scan clock and therefore make the rest of the scan data inaccessible. It was also found that sometimes two separate scan rings are needed, but only one can be active to ensure that additional test capabilities on the board are executable when the scan clock is disabled as allowed by the standard.

Also, as we pointed out in Section 5.8.4, if tristate–disable nets exist, then the pseudorandom tests generated may cause conflicts between the tristate–disable input and the outputs they control, when both of them lie on the same scan path. The relationship between the tristate control signal and the output must be deterministic but not random.

5.9.3 Summary

As compared with traditional methods, the boundary scan technique provides a more structured method for prototyping as well as product testing and debugging. This method allows a building block approach whereby individual circuits are first verified and then used to verify the interconnects between adjacent boundary scan ICS and finally the remaining portion of the board. This new approach to design, testing, and fault localization of complex digital systems creates a new process that is beneficial to both the design engineer and the test engineer because of the benefit of reduced time and efforts.

The boundary scan techniques described in the last few sections provide a real

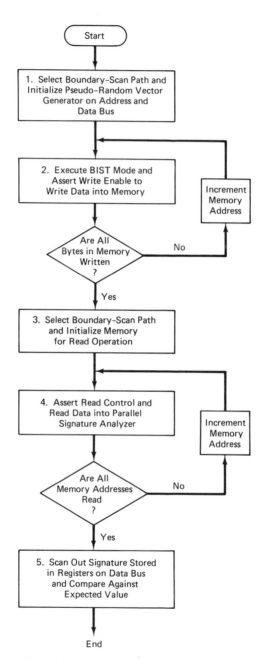

Figure 5.41 Logic flow of the BIST controller FSM.

solution to the problem of testing complex boards with high-density surface mount technology (SMT) devices. No doubt industry will jump on the bandwagon of boundary scan primarily for economical reasons. The end result will be wide applications of boundary scan technique and better tested products. Another added benefit of the advent of the boundary scan technology is the viability of using low-cost testing equipment in testing complex digital systems. As a result, the ATE industry itself will have to go through a transition from offering high-priced multipurpose testers to offering low-cost, microprocessor-based, single-purpose test instruments.

References

1. Ibarra, O.H., and S.K. Sahni, Polynomially complete fault detection problems, *IEEE Transactions on Computers*, **C-24**(3), 1975.
2. Bardell P.H., and W.H. McAnney, A view from the trenches: production testing of a family of multichip logic modules, *Proceedings of 11th International Symposium on Fault-Tolerant Computing*, June 1981.
3. Wang, Francis, BIST using pseudorandom test vectors and signature analysis, *Proceedings of 1988 IEEE Custom Integrated Circuits Conference*, May 1988.
4. Sedmak, R.M., Design for self-verification: an approach for dealing with testability problems in VLSI-based designs, *Proceedings of International Test Conference*, 1979.
5. Bhausar, D.K., and R.W. Heckelman, Self-testing by polynomial division, *Proceedings of International Test Conference*, 1981.
6. Lyons, N.P., FAULTRACK: universal fault isolation procedure for digital logic, *IEEE Interconnection Technology Program*, March 1974.
7. Savir, J., Syndrome-testable design of combinational circuits, *IEEE Transactions on Computers*, June 1980.
8. Frohwerk, R.A., Signature analysis: a new digital field service method, *Hewlett-Packard Journal*, **28**(5), 1977.
9. Savir, J., G. Ditlow, and P.H. Bardell, Random pattern testability, digest of papers, *13th Annual International Symposium on Fault-Tolerant Computing*, June 1983.
10. Agrawal, V.D., Sampling techniques for determining fault coverage in LSI circuits, *Journal of Digital Systems*, **V**, November 1981.
11. Stahnke, Wayne, Primitive binary polynomials, *Mathematics of Computation*, **27**(124), 1973.
12. Savir, Jacob, Probabilistic test, *IEEE Built-in Test-Concepts and Techniques Workshop*, October 1983.
13. Abramovici, Miron, P.R. Menon, and David T. Miller, Critical path tracing: an alternative to fault simulation, *IEEE Design & Test of Computers*, February 1984.
14. Jain, S.K., and V.D. Agrawal, STAFAN: an alternative to fault simulation, *Proceedings of 21st Design Automation Conference*, June 1984.
15. Smith, James E., Measures of the effectiveness of fault signature analysis, *IEEE Transactions on Computers*, **C-29**(6), 1980.
16. Konemann, Bernd, Joachim Mucha, and Gunther Zwiehoff, Built-in logic block observation techniques, *Proceedings of 1979 International Test Conference*, October 1979.
17. Katoozi, Mehdi, and Mani Soma, A BIST design of structured arrays with fault-tolerant layout, *Proceedings of 1988 International Test Conference*, September 1988.

18. Bardell, Paul, and William H. McAnney, Self-testing of multichip logic modules, *Proceedings of 1982 International Test Conference,* November 1982.
19. "JTAG Boundary-Scan Architecture Standard Proposal, Version 2.0," Technical Sub-committee of the Joint Test Action Group, Ipswitch, United Kingdom, March, 1988.
20. "IEEE Standard 1149.1, Standard Test Access Port and Boundary-Scan Architecture, Draft D3," Test Technology Technical Committee of IEEE Computer Society, New York, January, 1989.
21. Gloster, Clay S., Jr., and Franc Brglez, Boundary scan with cellular-based built-in self-test, *Proceedings of 1988 International Test Conference,* September 1988.
22. Hansen, Peter, Testing conventional logic and memory clusters using boundary scan devices as virtual ATE channels, *Proceedings of 1989 International Test Conference,* August 1989.
23. Wagner, Paul T., Interconnect testing with boundary scan, *Proceedings of 1987 International Test Conference,* September 1987.
24. Hassan, Abu, Janusz Rajski, and Vinod K. Agarwal, Testing and diagnosis of interconnects using boundary scan architecture, *Proceedings of 1988 International Test Conference,* September 1988.
25. Jarwala, Najmi, and Chi W. Yau, A new framework for analyzing test generation and diagnosis algorithms for wiring interconnects, *Proceedings of 1989 International Test Conference,* August 1989.
26. Goel, P., and M.T. McMahon, Electronic chip-in-place test, *Proceedings of 1982 International Test Conference,* 1982.
27. Vining, Sue, Tradeoff decisions made for P1149.1 controller design, *Proceedings of 1989 International Test Conference,* August 1989.
28. Halliday, Andy, Greg Young, and Al Crouch, Prototype testing simplified by scannable buffers and latches, *Proceedings of 1989 International Test Conference,* August 1989.
29. Wang, F.C., G. Nurie, and M. Brashler, An integrated design for testability system, *Proceedings of IEEE International Conference on Computer-Aided Design,* November 1985.
30. McClean, Don, and Javier Romeu, Design for testability with JTAG test methods, *Electronic Design,* June 1989.

Chapter 6

ATE and the Testing Process

Digital circuits are getting faster, denser, and more complex. The volume of digital products is also rising as price per function drops. And the users of these products are demanding more quality and reliability in the components or systems they purchase. Together, these trends are changing the nature of the automatic test equipment (ATE) industry. It's getting tougher—today it takes accuracy, capacity, flexibility, and economy for testing products to succeed in the fierce competitive market. No matter what type of testing or verification it is, more than ever, the best and the most cost-effective ATE is needed.

In this chapter, we will look at the digital circuit testing process as a whole, and we will discuss some important types of tests in terms of the features that need to be tested, but from a tester-independent point of view. We will then discuss the various types of testers that perform these tests. We will also discuss the topic of linking design and test. This is a link that both the CAE and ATE industry have been striving to establish for the last few years. We will discuss the kind of accomplishments that have been made in this area.

However, mixed-signal testing of analog and digital circuits, a concern with ever-increasing importance in industry today, will be deferred until the next chapter. This is an area involving more than just ATE, and therefore it deserves a place in Chapter 7, "Special Testing Topics and Conclusions." For similar reasons, microprocessor testing will also be discussed in the next chapter.

6.1 The Types of IC Testing

The testing process in a broad sense encompasses all the quality assurance activities of a product, including incoming parts inspection, prototype verification, new device characterizations, IC process monitoring, and circuit and board testing. Although each testing activity has its focus, their overall objective is to ensure the quality and reliability of the end product.

In this section we will discuss the types of tests normally required for digital circuits and systems. Most of the materials in this and next section are adapted from product descriptions[1] of Integrated Measurement Systems, Inc. Each type of test must be performed in the most efficient manner so that its contribution to the total cost of the end product can be minimized. The tests to be discussed below are classified as functional test, AC timing test, and DC parametric test. They are performed in almost all the phases of the testing process.

6.1.1 Functional Tests

The primary objective of functional testing is to verify the logic of the circuit to see if it performs according to design requirements. The process of functional tests involves applying test stimuli to the input pins of a circuit or a board under test and measuring its output in order to compare with the expected output. The source of the test stimuli is frequently the simulation vectors used by the design engineer to verify the design, but at times these test sets are augmented by automatically generated test vectors. The number of test vectors needed for functional testing can be quite large.

Since functional testing is to verify that certain inputs result in expected outputs, other parameters such as circuit timing, input driving levels, and output threshold levels are set to conservative levels. Timing requirements still need to be verified in this type of testing, even though they are not pushed to the limits as in AC timing tests.

The simplest functional testing is a go/no-go type of test. In more complicated cases, especially in verifying a prototype, a more involved process to isolate the fault is part of the functional test. This occurs if the test responses do not match the expected outputs and the circuit is considered as having failed the functional test. To pinpoint the cause of the failure, the circuit goes into the diagnostic mode, which frequently requires the generation of a lot more tests. The key requirements here are flexibility and interactivity so that the test engineer or design engineer can easily and quickly make changes to the test data, apply these tests, and observe their results.

The strategy of partitioning a failed circuit into verifiable sections usually is a step that helps the fault isolation process. The testing equipment must support these features in order to be useful. Other design-for-testability features such as scan-path and boundary scan have unique requirements for testers that use simple read/write pattern buffer memory. For example, the pattern source may need to supply multiple channels of serial data for circuits under test (CUT) with multiple scan paths. Furthermore, the test engineer must have a way to designate scan sequences of specified lengths. If the CUT has a scan-path of 50 elements (i.e., scan registers), for instance, the tester must be able to initiate a scan sequence of precisely that length. So each scan channel must have a way of being subdivided into modules of, in this case, 50 elements.

To adequately test serial scan-designed circuits, a deep serial memory for storing the test vectors is needed. When multiple serial shift chains are needed, the scan registers must be easily reconfigurable up to 16 or more chains. To diagnose a fault, the test engineer must be able to partition the serial stimulus into segments in order to apply them at different times throughout functional testing to the various diagnostic pins. The tester also needs special facilities to trace and store scan errors. Errors should be traceable to a specific failing bit that is distinguishable from all other failing bits—pinpointed by channel, scan-vector, and bit.

6.1.2 AC Timing Tests

The primary objective of an AC timing test is to determine if the CUT satisfies various timing specifications[2] so that its behavior can be predicted when it is placed on a board. Examples of timing measurements that are made in an AC timing test include rise time, fall time, pulse width, pulse aberrations, setup time, hold time, and propagation delay. Delay measurements between level crossings at any pair of pins or from any reference start to any stop pin are also important in this type of test.

These measurements are often performed under varying conditions to check the response over its stated operating ranges. Often, measurements are made running multiple passes of a test while varying one or more parameters. As a result, this type of test is also referred to as an AC parametric test. Normally, maximum and minimum operating temperatures and tolerances for DC power inputs are used to characterize the safe operating regions of the circuit. Finally, the variation in response characteristics under repetitive stimulation or longrange operation is checked.

As far as ATE requirements are concerned, timing accuracy and resolution, data formatting, and programmable timing are essential for AC timing tests. When a large tester is testing a high-speed CUT that operates in the nanosecond range, signal path timing skew is a very significant problem. Signals must arrive at the CUT with the proper timing relationship to achieve meaningful results. The tester must have a way for automatic deskewing for superior timing accuracy. A resulting resolution in the range of picoseconds should be achieved. The tester must also be deskewed whenever calibration is performed so that longterm timing drift is stabilized.

Setup time, defined in Chapter 1, is measured[1] by delaying the leading edge of an input pin's signal transition incrementally until it causes a failing in a functional test vector. The difference in the data delay for the input pin's signal and the delay for the clock signal gives the setup time T_{st}. Figure 6.1 shows the setup time between the data input pin and the leading edge of the clock. In actual

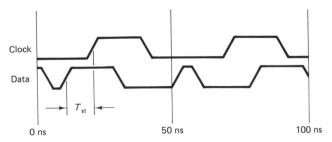

Figure 6.1 Setup time measurement.

measurements, multiple tests are run with increasing delay until the test fails due
to the data input setup time being violated.

The hold time, also defined in Chapter 1, is measured in a similar manner
except that the trailing edge of an input pin's signal transition is reduced until it
causes such a failure. Figure 6.2 shows the hold time T_{ht} as the difference in
delay between the data signal for the input pin and the clock signal. Several data
formats can be used to program the input pin to be measured, including return-
to-complement (RC) format used in Figures 6.1 and 6.2 for the data signal.
Delayed non-return-to-zero (DNRZ) format can also be used in setup and hold
time measurements, because both RC and DNRZ formats allow a 0 to 1 and a 1
to 0 transition to occur in a given cycle and therefore they provide the rising edge
and falling edge needed for the measurement.

Propagation delay T_{pd} for a signal from an input pin to the output pin of interest
can be measured easily by using the RC format for data input. It is determined
by measuring the leading edge of the clock and the transition of the output pin
as shown in Figure 6.3.

Various data formats are required for AC timing and other tests. Some of the
commonly used formats are shown in terms of sample waveforms in Figure 6.4.
These data formats basically specify the various waveforms that can be applied
to a circuit's input pins. But the actual placement of the waveform edges are

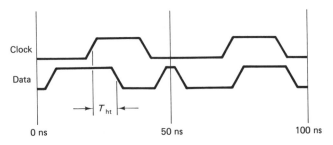

Figure 6.2 Hold time measurement.

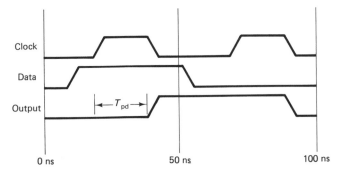

Figure 6.3 Propagation delay measurement.

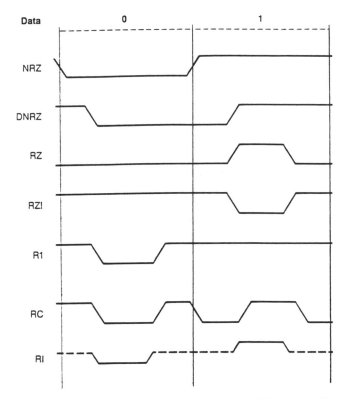

Figure 6.4 Data formats needed for testing. (Courtesy of Integrated Measurement Systems, Inc.)

determined by the occurrence of the clocking signal that is applied to drive the data into the circuit. The various data formats are defined[1] below.

- *Non-return-to-zero (NRZ):* A format that transitions only at the clock cycle boundary. It is the simplest format and is frequently used for the data input bus.
- *Delayed non-return-to-zero:* A format exactly like the NRZ format except that its waveform is shifted from the cycle boundary by a user-specified delay value. It is also commonly used for the data input bus.
- *Return-to-zero (RZ):* A format that represents a waveform at a low logic level at the beginning of the clock cycle. It transitions to a high logic level for a user-specified pulse width and then return to a low logic level. It is frequently used for the clock signal.
- *Return-to-zero inverted (RZI):* A format that is simply the inversion of the RZ format. It is frequently used for the control signal.
- *Return-to-1 (R1):* A format that is similar to RZI except it pulses low when the data value is 0, while RZI pulses on the data value of 1.
- *Return-to-complement:* A format that specifies a logic value and a time interval during which the data remains at this logic level in each clock cycle. Outside this interval, the complement of the specified logic level is maintained. It is frequently used in setup and hold time measurements.
- *Return-to-inhibit (RI):* A format that specifies a data value on a tristate bus. It is in a high-impedance state at the beginning of a clock cycle and then transitions to a user-specified logic value before returning to the same high-impedance state. The time interval in each state is user programmable. It is frequently used in verifying a bidirectional bus so that no input signal is applied at the cycle boundary when the bus is being switched from input to output mode.

Other measurements of interest in an AC timing test are the maximum operating speed of the CUT and the detection of glitches caused by internal race conditions. The latter can be detected by sweeping the sample time across the clock transition time to uncover short output pulses occurring at the wrong place on the timing scale, as shown in Figure 6.5.

6.1.3 DC Parametric Tests

The primary objective of a DC parametric test is to determine a CUT's DC driving characteristics of its output and the loading characteristics of its input. This type of test is as important as functional and AC timing tests because if a circuit does not satisfy the prespecified limits of its DC parameters, then it will fail on a board or cause other parts on the board to fail when such limits are exceeded. Important DC characteristics to be tested should include stress, leakage, breakdown, resistance, continuity, opens, shorts, differential voltages,

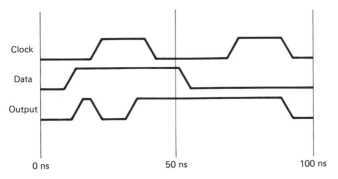

Figure 6.5 Glitch detection.

power supply connections, current under forced voltage, and voltage under forced current. The last two measurements are concerned with the driving capability of a CUT and its DC loading.

To determine input pin DC loading, a given voltage is applied to the CUT's input pins while the current flowing into the circuit is measured. Usually a range of DC voltage is applied to the input pin, and the range should include the value for a logic level high and the value for a logic level low. From this data, it can be determined if the circuit's input pins present too much load to the circuit driving it on the board. If indeed they do, then the input levels will be pulled down and the CUT will fail when placed on the board. For this type of measurement, the CUT sometimes has to be initialized into a known state before testing. This requires additional vectors to be applied first to initialize the circuit. This step is frequently called functional preconditioning. It is also noted that sometimes a given current is applied to the input pins while the voltage across the circuit is measured. This type of loading characteristic is also important if the driving circuitry on the board is a constant-current or constant-voltage device.

To determine the output pin DC driving characteristics, a similar step is taken as in the input pin loading measurement. For each output pin, a specified current is fed into the pin to present a DC load, then the voltage across it is measured. As in the case of the input loading measurement, a range of output voltage is measured which should include the output voltages corresponding to a logic level high and to a logic level low. From this data, it can be determined if the CUT's output pins provide adequate DC drive into the circuitry on the board. If not, then the DC level may be pulled down and the board may fail when the CUT is placed on the board. Sometimes it is also necessary to apply a logic high or logic low voltage to an output pin and then measure its current drive characteristics. Again the CUT sometimes has to be initialized prior to making DC parametric measurements.

In DC parametric tests, programmable voltage supplies are used by ATE as the sources for the different forcing voltages. These must be of various ranges to satisfy the different measurements required. Additionally, current-forcing supplies with full-scale range are also needed. The resolution of these supplies should be orders of magnitude less than 1% over their ranges. Overall, these DC measurements must be accurate and fast. They are time-consuming if done manually because of the many steps involved. The ATE must provide the test engineer a detailed summary report that lists all the test results on a per pin basis so that troubleshooting can be done easily.

6.2   Modern Semiconductor Testers

In the early days of electronic industry, the testing systems or equipment were mostly the in-house type to suit a particular company's internal needs. They were primitive compared with today's modern ATE with powerful microprocessor and sophisticated testing fixtures. However, the electronic industry was also not as advanced as it is today, and whatever was available served its purpose well. There were very few general purpose ATE's in those days. Computers were also not used extensively in the testing equipment because the computer industry was in its infancy itself.

The advent of microelectronics changed all this.[3] The ever-growing complexity (speed as well as density of integrated circuits and the pervasive usage of digital electronics in products created many challenging testing problems to solve) therefore stimulated the ATE industry to grow. Two trends seem to be obvious today. The first trend is the continued exploitation of microelectronic technology by the ATE industry for faster, more accurate, more economical, and more compact testing systems. The second trend is the pervasiveness of digital computers in both semiconductor and board-level testing systems not only to control and sequence the testing process, but also to provide a software development environment to create and maintain the associated test programs and test libraries.

In this section, we will look at testers used primarily in testing semiconductor devices. We will discuss the architecture of a modern ATE in terms of its major subsystems, and we will also discuss various technical considerations and requirements of a modern ATE.

6.2.1 Architecture of a Testing System

A semiconductor tester's primary function is to test a stand-alone integrated circuit such as a microprocessor, a memory, or a random logic device. A high-performance semiconductor tester can perform functional, AC timing, and DC parametric tests. A modern semiconductor tester[4] normally consists of a powerful digital computer together with the necessary peripheral devices, per pin elec-

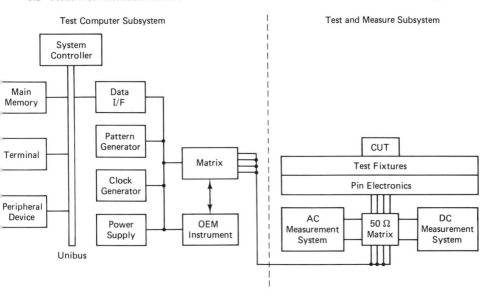

Figure 6.6 Block diagram for a modern ATE.

tronic circuitry for driving and monitoring the input and output pins for the CUT during different types of measurements, and loads and test features for connecting the CUT to the tester. A block diagram for a typical modern semiconductor tester is shown in Figure 6.6. The basic ATE structure is made up of the test computer subsystem and the test and measure subsystem. The test computer subsystem contains all the computing power and equipment to automate all the testing and related functions. The test and measure subsystem contains all the measurement electronics for the CUT.

Now we look at each of the key components of the test computer subsystem in Figure 6.6.

- *System controller:* A fast and powerful processing unit which controls the testing system through a bus structure. It executes the test engineer's test program by transmitting data and receives subunits' status reports, interrupt requests, and test data. It is frequently a commercial microprocessor, but it must be compatible with the system's test and measurement capability.
- *Main memory:* The main memory associated with the test computer. This must be supplemented by additional local memory for each I/O pin for storage of per pin test data.
- *Peripheral device:* A set of equipment in support of the CPU to meet the requirements of the testing environment. Typical peripheral devices include disks, magnetic tapes, line printers, graphics terminals, and hard copy units.

- *Pattern generator:* A processor which controls all test pattern generation and sequencing functions. It also controls the clock generator to select the proper timing sets and run mode. It uses loops, subroutines, and list pointers to generate repetitive patterns or to modify patterns presented at each cycle of test. All functional data for the four test modes (force, compare, inhibit, and mask which will be explained in the next subsection) are supplied by the pattern generator. Note, however, that this is a not ATVG itself because these generated patterns are based on the test program flow not on the CUT's internal structure and fault models commonly used by an ATVG.
- *Clock generator:* A multiset generator of multiphased clocking signals. These sets of timing data are programmable and selectable on a cycle-by-cycle basis. Cycles should be generated in a free-running mode or by synchronization of the CUT. This mode should also be switchable on a cycle-by-cycle basis.
- *Power supply:* A set of programmable power supplies with programmable current limits to protect the CUT from supply current reaching destructive levels. Both current-forcing and voltage supplies with range resolution are needed.
- *Matrix:* An analog switching matrix to allow additional OEM instruments to be connected to the test station through the 50 Ω matrix. This expands the number of signal paths in the system. Under programmable control, any specific external equipment can be interconnected with the tester.
- *Data I/F:* A general interface between the test station and test controller and the rest of the testing system. This enables information exchanges of test data or test commands. An IEEE Standard 488 bus interface[2] should also be allowed.

The key components of the test and measure subsystem are defined below.

- *Pin electronics:* The hardware that provides universal capability at each CUT pin. Under programmable control, each pin can be assigned as input driver, output comparator, bias supply, or ground. Many of the test and measurement features such as programmable drivers, programmable comparators and output loads, serial pattern/error buffers, and control logic are built into this circuitry. Under programmable control, drive high or drive low voltages can be applied to each pin.
- *Test fixtures:* The wiring socket circuitry that allows each pin of the CUT to be connected to the proper driver, receiver, I/O, power, or ground. These fixtures must contribute to minimum distortion or timing skew for signals transferred through them. They must also be compatible with all device package types such as dual-in-line package (DIP), pin-grid array, and leadless carriers. Interfacing with a device handler for incoming inspection and fixturing to a die for wafer probing are also required.

- 50 Ω matrix: A multichannel switching matrix that provides bidirectional access to all tester pins and impedance matching from the subsystem's pin electronics all the way to the device socket. Channels in the matrix are assigned for AC and DC parametric testing.
- *AC measurement system:* Contains the logic that controls the pin electronics to perform timing measurements, while itself is controlled by the system controller in the test computer. It performs such functions as determining when to apply test stimuli and when to strobe the output responses, placement of leading and trailing edges of data waveforms in timing control, automatic deskewing for timing accuracy, and time interval measurements for pin-to-pin mode and for reference-to-pin mode.
- *DC measurement system:* Contains the logic that controls the pin electronics to perform DC parametric measurements, while itself is controlled by the system controller in the test computer. It performs such functions as control of measurement repetition rates for high throughput, conducting parallel go/no-go DC tests on all pertinent pins simultaneously, automatic ranging for best resolution in measured results, and interfacing with the switching matrix for proper load balance.

Note that it is almost universally true that a tester must be modular in structure because modularity provides cost-effective upgrades as new functionality is needed. Frequently, the pin electronic card is the fundamental building block and a testing system must therefore offer practical configurability of pin count to meet evolving needs. Pin count should be added at no degradation of system performance and without disproportionate cost. Easy field upgrade from simple systems should be allowed so that a customer can start with a minimum configuration to satisfy initial requirements to reduce up-front capital costs. With the reserve of expendability, modules can be added easily when future needs arise. Other components such as power supplies, local memory, clock generators, and peripheral devices should also be modularized.

Other system requirements include flexibility in architecture so that a tester can work with other equipment, workstations, a sound software development environment, and interfaces with CAD/CAE software.

6.2.2 The Testing Process

To determine the most cost-effective way of testing a semiconductor device, a testing strategy must first be established. This strategy should prescribe a testing approach that is consistent with the quality standards in use. It must include such factors as types of tests, testing resources needed, overall accuracy requirements, and sources of test stimuli. Many alternatives must be considered and decisions have to be made. All of these require difficult tradeoffs to be chosen. One of the most important considerations in this phase of establishing a testing strategy is the accuracy requirement for the various measurements. For instance, a test engineer must choose among several approaches to compensate for the inherent inaccuracies in a testing system. When testing mature and high-yield process, the test engineer may choose *spec testing,* that is, to test the CUT exactly at rated specifications and rely on statistical probability to counteract the limitations of a given testing system's lack of accuracy. But for a semiconductor process that is not completely mature, process drift can occur and be hidden by the tester's tolerance. The test engineer can therefore choose an alternative called a *guard-band approach,* that is, to use a window established by adding the tester's worst-case tolerance to the device specification to reject a CUT if it falls outside the window. Thus marginal parts or bad parts will not pass the test. This approach is used with all but the most mature semiconductor process and technology where it can reduce the overall yield. A possible strategy is to reduce the guardband area without sacrificing any quality objective.

Once all the tradeoffs and decisions are made, a test plan is produced which documents all the testing requirements regarding the circuit to be tested. Such a document is used as a high-level guide for all subsequent testing activities. But it must first be approved by the quality assurance committee involving the design, manufacturing, marketing, and the test departments.

Based on the approved test plan, the test engineer can start to prepare a test program in a testing language such as the IEEE Standard ATLAS test language or in a system-specific test language. The test program contains the executable codes to run on the testing computer and is the major output for the test engineer. Test vectors generated either by an ATVG or manually for design verification are also included in the test program. The code in the test program specifies line-by-line all the steps necessary to test a circuit by the test and measure subsystem. These steps involve setting up the configurations of the CUT, the levels and resolutions of the pin drivers, measurements to be taken, values to be compared, timing to be applied, and output to be printed. The test program is also used as a document that specifies in great detail all the tests to be performed on the CUT. It frequently serves as a vehicle of communication between the test engineer and the design engineer, especially when the test uncovers problems.

Actual testing starts after the CUT is fixtured in the tester and the test program is ready to run on the testing computer. A sequence of commands are sent to the various components in the two subsystems to initiate testing actions. To perform the necessary tests, the pin driver circuit can be put into any of the four test functions on a cycle-by-cycle basis: force, compare, inhibit, and mask. For the force function, the pin electronics drive the CUT pin as an input channel, while for the inhibit function the pin electronics are isolated from the CUT when its driver is put into a tristate mode to test a bus-structured device. The compare and mask functions define when during any clock cycle the test data should be compared and when it should be ignored.

During testing, the pattern generator provides all the test patterns to all the pin

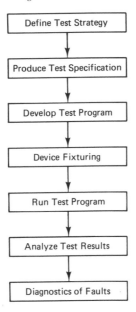

Figure 6.7 The testing process.

electronics cards. The multiphase system clock provides synchronization of test functions applied to each CUT pin. During each clock cycle, the test program also selects force, compare, inhibit, and mask functions for each pin. The function of when to apply test stimuli and when to strobe the output is controlled by the AC measurement subsystem. Error detection is done by the compare function and failure data is logged into the system's local memory. The incorrect data and the related address information are captured whenever an error occurs. The stored errors are then used as a mask source during a second pass through the test, so that only new errors are recorded. These test results are analyzed later and used for diagnostics and troubleshooting by the test engineer.

An overview of the testing process just described above is shown in Figure 6.7.

6.3 Board-Level Simulation and Testing

6.3.1 Board-Level Testing

All semiconductor devices once tested and verified will have to be placed on a printed circuit board in order to form a part of the digital circuit/system of a product. Modern computer-aided manufacturing (CAM) technology has automated the whole process from fabrication of the board through assembly of components on the board until it is ready for testing. CAM uses auto-insertion machines to place through-hole devices on the board, and it uses pick-and-place machines to mount surfaced-mounted devices (SMDs). Although CAM technology facilitates the manufacturing process, it does bring additional problems for testing board-level assembly of a digital system. Physical defects such as a bent pin or incorrect parts can be introduced by the associated automation equipment and processes. To deal with these problems, additional testing steps are introduced.

During manufacturing of the boards, bare-board testers must check continuity before assembly takes place. After assembly, in-circuit testers are used to verify individual components and circuits for correct placement on the board and correct electrical value. Functional board testers access the various edge connectors and test the functionality and interaction of parts on the board as a unit.

Test vector generation for functional board testers is greatly simplified if test vectors are available from the CAE system's board-level simulators. As a result, linking manufacturing with the design process is necessary in order to allow critical data from CAE and CAD systems to be transferred into the computerized factory floor. We will discuss board-level simulation in the next subsection.

The board-level testing process is illustrated in Figure 6.8. Note that the whole manufacturing and design process has interfaces with a manufacturing resource planning (MRP) system supported by a corporate-wide MRP data base. The primary function of bare-board testing is to check continuity between any pair of conducting points and also to check the associated leakage resistance. It frequently employs a known good board and then compares all other boards fabricated with it to detect any manufacturing problems.

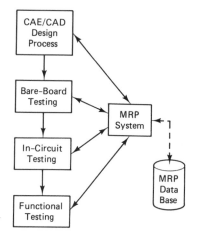

Figure 6.8 The board-level testing process.

One of the difficulties encountered in board-level testing is the derivation of functional test vectors for the whole board. It is not like IC testing, in which design verification vectors can be used as a basis for functional testing. There are seldom any design verification vectors from the board-level simulation environment to use in functional testing of the whole board. But the trend has changed in the last few years toward more board-level simulation and less prototyping for verification and debugging of complex printed circuit boards. This is made possible by the availability of functional level simulation models which incorporate both behavioral and timing features of complex digital devices like microprocessors. The impact is not merely in the design verification phase of a printed circuit board but also in the testing phase as well.

6.3.2 Board-Level Simulation

Various simulation techniques have been used extensively by IC designers in the last decade to verify the design of a circuit's functionalities as well as its timing characteristics. These techniques have been proven extremely useful in cutting down the design time and the number of iterations in the manufacturing of the integrated circuit. The simulation process is frequently called software prototyping, and it has completely replaced prototyping of a digital circuit by hardware because an IC does not allow hardware prototyping easily. On the contrary, hardware breadboarding had been regarded as an efficient way to prototype a printed circuit board until only a few years ago. Normally the prototyping phase of a board involves five or more steps from design, layout, fabrication, assembly, all the way to testing. This is a lengthy process and the associated design development cost can be staggering if many iterations are necessary to debug a board.

Aside from economic considerations, breadboarding has been made obsolete as a way for prototyping by new technology (e.g., surface mount technology and double-sided boards) used in board manufacturing and the increased complexity and density of a typical digital board. These trends made the conventional method of prototyping less attractive from the point of view of cost and design time. The only alternative is to use simulation as a way for prototyping, as in the case of integrated circuits. Meanwhile, a few CAE tool manufacturers started to produce software models of complex digital circuits and devices. These models are functional behavioral simulation models that contain[5] intelligence of the devices modeled such as timing requirements, built-in error checking, handling of unknowns, and preloadable memories to enhance simulation. These changes together with the advancements made in simulation techniques and the availability of powerful workstations have now made board-level simulation a viable and necessary step in design verification of printed circuit boards.

The materials used in this section are adapted from product descriptions[5] from Logic Automation, Inc. As shown in Figure 6.9, the number of steps in the prototyping phase is reduced to two through software prototyping. Besides func-

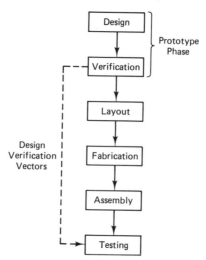

Figure 6.9 Linking board simulation with test.

tional verification, timing analysis is also possible with board-level simulation. This provides data for both propagation delays between clocked elements as well as worst-case delays. The latter is achieved by varying combinations of minimum and maximum delays of devices during simulation.

The use of board-level simulation for design verification has also benefitted board-level testing because of the availability of the simulation vectors for use in testing. The software models used in simulation focus on the functionality of complex devices such as a microprocessor at its bus cycles level, and therefore hardware verification is possible. Being closely linked with popular simulators, these models also provide the simulator users with a window to view the inside of the models during simulation. With this ability, the designer will know what is in the device's registers as illustrated in Figure 6.10. Using the simulator, microprocessor design verification tasks such as tracing instruction execution and setting break points can be carried out.

6.4   In-Circuit Testers

Since bare-board testing is relatively simple compared to in-circuit and functional testing of a printed circuit board, we will focus our discussions on board-level in-circuit and functional testers in this section and the next section.

The primary function of in-circuit testing of a board is to detect any defects caused by the manufacturing process for each part on the board. This type of test is performed through bed-of-nail or probe contact to each circuit node on the

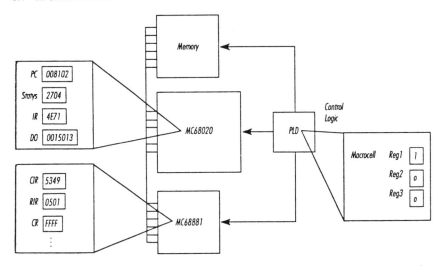

Figure 6.10 A window into key devices during board simulation. (Courtesy of Logic Automation
Incorporated.)

board.[6] These nails allow access to every node on the printed circuit board so
that individual circuits can be controlled and tested. The board under test is
placed on the nail-bed, which is normally vacuum-activated during testing to
ensure good contact between the spring-loaded pins that connect the ATE electronics with electrical points on the underside of the board. A typical test fixture
is shown in Figure 6.11.

   In-circuit testing requires isolation of each circuit on the board so that each

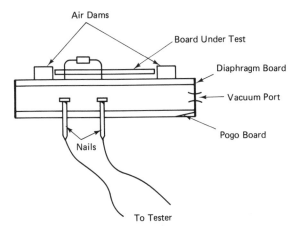

Figure 6.11 A bed-of-nail test fixture.

can be individually tested without the influence of surrounding parts. This is achieved by controlling circuits adjacent to the CUT so that their states do not affect the CUT. Adjacent circuits are usually put into a tristate since its output is floating in this case. If tristate is not possible, the tester attempts to backdrive the part by forcing it to a target state. Although the current required to backdrive a part can damage it if applied for too long,[7] the few milliseconds required by the tester generally will not damage the part. As a precaution, the testing should proceed from the output of the board toward the input so that a part damaged by backdriving in one test can be detected in subsequent tests. Isolating each circuit during testing is important because unless it is free from all stimuli from the rest of the board, the tester can not adequately test the function of the part.

In-circuit testing is distinct from functional testing of a board, which applies test stimuli to the input edge connectors of the board and monitors its responses at the output edge connectors instead of accessing input and output pins of individual circuits internal to the board. If a board fails during functional testing, there is no way to tell which part is causing the failure. Since each part is tested separately with in-circuit testing, a failure can be isolated to a specific part.

An in-circuit test is also usually run at a rather low test rate[8] because at a high test rate the test becomes more sensitive to timing delays and waveform inaccuracies which in turn make the test very hard to debug when a failure occurs. However, in functional testing, it is desirable to run at the maximum test rate possible so that a real operational environment of the board is simulated in order to detect all potential failures and defects.

During testing, the in-circuit tester applies overdrive voltage/current to the nails connected to the CUT's input pins while monitoring the responses on the nails connected to the output pins. In-circuit testing can detect such physical defects as bent pins, broken pins, incorrect parts, and parts inserted with incorrect orientation; it can also detect internal faults of the CUT through functional tests. But in-circuit testing is seldom used to check a circuit's timing parameters, such as propagation delays, maximum operational speed, or DC parameters.

The test vectors used in in-circuit testing normally come from three different sources: functional verification vectors from the design engineer, fault detection vectors from ATVG, and auxiliary vectors from ATVG or from the test engineer. In-circuit testers use auxiliary vectors called inhibit, disable, H_{force}, and L_{force} to isolate parts other than the CUT, to perform bus tests, and to perform tests on the CUT. Both the disable and inhibit vectors attempt to keep parts from changing state during the test in order to test the CUT that is either electrically adjacent or on the same bus. Usually a bussed circuit has one or more tristate enable pins that can be used to set its output pins to a high-impedance state, effectively disconnecting them. A typical test condition is shown in Figure 6.12, where one IC is disabled while the other is inhibited to keep them from interfering with the

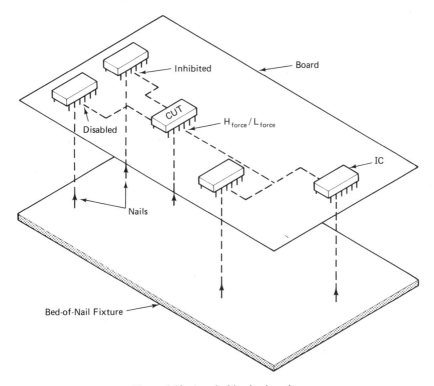

Figure 6.12 A typical in-circuit testing.

testing of the CUT. Note that the CUT's output pins are driven to a logic high or a logic low level by an H_{force} or L_{force} vector(s), since they are connected to a bus in this test. Also shown in this figure are the bed-of-nail fixtures for setup contacts with the proper electrical nodes. These connections are shown as dashed lines because they may not be visible due to multilayered board structures.

A circuit can not be disabled if it does not have tristate control lines or if it is tied to power or ground. The circuit may toggle its output due to feedback loops or free-running clocks. In this case, other means must be used to keep the circuit in a stable state to prevent it from interfering with other circuits during testing. For a TTL device, backdriving it to a logic high level is a viable alternative.

The H_{force} and L_{force} vectors force the output pins of a CUT into a logic high or logic low value during bus test. For instance, in testing a microprocessor the address bus, the read line, and the I/O line must be forced low during bus test.

Most in-circuit testers use libraries of tests and test procedures for each of the circuits on a printed circuit board to isolate and test them. For standard TTL parts, the test engineer creates a test procedure, stores it in the test library, and reuses it every time the part is tested. However, since PLDs can be programmed differently, and may react differently depending on their configuration, a PLD test procedure is not reusable.[9] The test library must be managed by a librarian function which supports creation, addition, deletion, and modification of test procedures. In summary, in-circuit testing uses both functional vectors and ATVG-produced vectors to test each part individually on the board to detect both physical and functional faults. Additional vectors are also needed to isolate parts on the board from interfering with the testing of the CUT. This type of testing is effective for debugging a board with through-hole devices, but it suffers from the difficulty of accessing SMDs. Therefore, it is not very effective in testing boards with many SMDs.

6.5 Functional Board Testers

6.5.1 Testing Philosophy

Unlike in-circuit testing, which clearly defines what type of tests are normally performed and where the test vectors come from, functional testing of a board itself does not have such a clear-cut definition. We can say that its objective is to test the whole board as a single system or unit to determine if it performs the prescribed functions. However, the types of tests that are needed to meet such an objective are not obvious. We need to examine the philosophy of board-level functional testing in order to arrive at some conclusion.

As the complexity of an IC increases, so does the difficulty and the cost of producing a completely comprehensive test. A full functional test of a complex part such as a 32-bit microprocessor may require millions of test vectors.[8] Instead of blindly verifying all possible combinations of inputs and outputs, a testing philosophy is necessary to guide the selection of tests. Rather than disregarding the inner workings of the device, the test engineer should try to use the knowledge of its functionality to produce tests that exercise the internal functional units individually. An approach like this can produce the type of tests that effectively verify the interworking of the components of the device without incurring the penalty of an exhaustive test.

The fundamental requirement for this kind of testing approach is the willingness of the test engineer to become thoroughly familiar with the device's logic, electrical, and timing characteristics. Microprocessors and similar devices require initialization to bring them up to a known state before actually exercising its functionality. Sometimes it requires the iteration of the same vector sequence many times before the device settles into an initialized state. Fortunately, most ATE today use pattern generators that can perform this function efficiently both in time and in memory. In addition, a processor instruction cycle must be synchronized to the tester. In many cases this can be done simply by assigning a test port to the clock pin of the microprocessor and synchronously stepping the processor through a series of instructions.

To start, the test engineer must study the manufacturer's data sheet and examine test procedures for similar devices to determine if the device is electrically testable. The tester's driver typically slews at about 30 V/μsec into 100 Ω resistive, a rate that is too slow for some IC inputs.[8] A test nail typically presents a 100–300 pF load to an IC output pin, which some devices will not be able to drive properly. These and other electrical requirements must be thoroughly understood by the test engineer in order to decide on a test approach.

Manufacturer's design specification should not always be accepted as test specifications. Many processors with dynamic registers can run considerably slower at room temperature than at their minimum clock frequency, which usually applies to the maximum operating temperature. The manufacturer may likewise have specified rigid timing or slew rates to ensure proper system operation rather than device operation, so that an adequate functional test of the device by itself can easily tolerate some timing limitations. Functional testing should not be used to verify a device's timing parameters.

6.5.2 Testing Objective

Since it is impractical to test all possible combinations of inputs and outputs, a functional testing should accomplish the following objectives.

- Toggle all pins high and low
- Exercise all major functional blocks
- Verify all addressing modes
- Verify all data types
- Access all internal registers
- Perform typical user instruction sequences
- Verify all data paths

Guides for designing a thorough test can be found in the block diagram of the device provided by the manufacturer and in the structure of the device commands. Often a functional or addressing mode will correspond directly to a bit in the instruction, so that a set of representative test instructions can be chosen by having each mode bit set and cleared at least once during the testing process. The device block diagram should also be analyzed to determine that each functional unit indicated on the diagram (e.g., ALU, registers, and counters) and each data path (e.g., from an accumulator to a memory address latch) is tested. Interrupts and status flags must both be tested and verified. The above guidelines should help a test engineer to develop a board-level functional test plan for a set of complex digital devices.

A most current but complementing approach to the guidelines discussed above is the use of board-level functional simulation vectors as test stimuli for board-level functional testing. Board-level simulation is made possible by the new modeling approach to complex devices,[5] as was discussed in Section 6.3. Complex devices such as microprocessors and their auxiliary devices are now made into software models representing the microprocessor's function at its bus cycle level. These models also include detailed timing specifications for each component within the model. Through a set of compilers that produce simulator-specific simulation models, these models can be used by many popular simuators. This approach has the advantage of representing a complex device's functionality at an appropriate level to verify its timing requirements and interactions with other devices in a board environment, without the need of a real part for the device to exist.

These functional models provide a way to simulate a board full of complex devices. They also solve the problem of the lacking of board-level ATVG today, because the functional simulation vectors can perform many of the functional tests to satisfy the objectives outlined above. Development of effective test vectors is the key problem in functional testing and sometimes it takes up to one-third of total design efforts.[5] With additional functional vectors developed for specific test requirements, the functional simulation vectors form the framework of comprehensive board-level functional test sets. With these software models, board-level fault simulation is also made possible to determine how effective the test vectors are at detecting all potential faults on the board.

Fault simulation also serves as a bridge between board simulation and test because fault dictionaries[10,11] can be generated for diagnostics at component level at the I/O pins of the board under test. Based on the assumption that an internal fault on a board will have its effect propagated to one or more of the output pins of the board, the fault simulator records the observed output state and the test vector(s) used. The fault dictionary is a giant data base produced by fault simulation that contains all the symptoms (i.e., observed output states) and the sources of the problems (i.e., the responsible internal faults). A test engineer armed with the information provided by a fault dictionary can identify the faulty components that are responsible for the observed test results. Proper repair actions can take place as a result. This is the major objective of fault diagnosis.

6.5.3 The Functional Testers

Board-level functional testers usually have architecture similar to that of semi-conductor testers. Although most board-level ATE combines both in-circuit and functional testing functions in the same system, the functional testing subsystems are different from the in-circuit testing subsystems in major ways.

First, the bed-of-nail type of test fixture for in-circuit testing can not be used for functional testing because it frequently fails at frequencies above 2 MHz due to interfering crosstalks.[12] In-circuit test fixtures sometimes take the form of a

twisted pair of wires, and therefore there is a large chance that the associated transmission line mismatch at high frequency can dramatically affect the performance of the test fixture. Because of these problems, functional testing seldom uses the same test fixture used by in-circuit testing.

Functional testing requires a high-speed test so that circuits requiring complex test timing or high minimum test rates can be efficiently tested. Therefore, the high-speed subsystem in the tester must provide a number of highly specialized capabilities for microprocessor testing. These capabilities are implemented either in hardware or software and must also be compatible with the standard digital test hardware. For one thing, the high-speed subsystem must exert great control over test timing parameters. While the test steps are programmed in the same way, the steps used to toggle clock pins and maintain tester synchronization can usually be eliminated from these high-speed functional tests. In their place, the clock generator subsystem will be used to establish steady-state clock signals, and the driving and sense timing is also set.

In addition to taking over the clocking function from the test program, the high-speed subsystem is also designed to simplify the testing of a microprocessor's irregular machine operations. A microprocessor operation sometimes extends over many clock cycles during which nothing of interest to the tester occurs, and it can also start or finish at an imprecisely known time, as in an asynchronous bus transaction, for instance. As a result, there are machine states that need to be tested only after a number of intermediate states have elapsed. During this time the testing activities can be suspended. This is done by including a trigger condition to suspend a test step. The step will not be executed until the required state is satisfied in the trigger condition. Thus, the functional tester does not fill steps or wait loops to keep itself in step with the microprocessor operations.

To summarize, functional testers have high-performance pin electronics and timing subsystems so that they are capable of emulating the end system. But they are limited by low channel capacity and lack of high-current drivers. Both in-circuit and functional testing capabilities are needed in a single board-level tester. A comprehensive test is possible for a complex board only if the strength of a high performance functional tester is combined with features such as high pin count and full nodal access from an in-circuit tester.

6.6 Linking Design and Test

Design automation technology today has matured to the extent that computer-aided design/engineering (CAD/CAE) tools are used extensively in all phases of product design. The trend in the testing area is also toward a fully automated testing environment called computer-aided testing (CAT). As a natural flow of the product from the phases of design to engineering to testing then to manufacturing, data used or produced in one area is often needed in other areas. For instance, the CAD data base which contains information regarding physical components on the board is needed by the test engineer for fixturing the board for testing.[13] Another example is the use of design verification stimuli and responses from CAE products such as logic and board simulators as a basis for the testing program. Other CAT tools such as ATVG, testability analyzers, or automatic test program generators must have a certain degree of access to the design and engineering data base, as well as to each other's data bases. All of these point to the importance of data exchange among these areas. Looking around in both the CAD/CAE and CAT industry nowadays, we can conclude that linking design with test is not just a trend now, but rather a way of life in both the design and test communities.

In this section, we will discuss the topic of linking design and test in terms of CAD/CAE data format translators and verification of stimuli/response compatibility. We will also discuss some key factors that are essential to make linking design and test successful.

6.6.1 Design-Test Link

Linking design and test does not merely mean the writing of a software program to translate from a design data format into that of a tester-specific format. Because of the multisimulator and multitester environment in most of the midsize or large companies present in industry today, cross-access and feedback links from test to design areas must be considered. Each of these often has its own formats and standards. Thus the design–test link is a multipath link involving a set of n by m format/standard translators. Each time a new format/standard is introduced into either the CAE/CAD or CAT environment, a new set of translators has to be developed. The resulting development and program maintenance costs can be prohibitive. A more severe problem is the management of such a large set of spaghetti-like translators.

To circumvent problems associated with this multipath linkage between design and test, companies that produce these links use an intermediate (neutral) format approach that reduces the scope of the problem from n (i.e, the number of simulators) by m (i.e., the number of testers) to that of $n + m$. Instead of translating each simulator format/standard into a target tester format/standard, it is translated into the neutral format first. Translators are also provided from the neutral format to that of a tester specific format/standard for each of the testers in use. Figure 6.13 shows a set of bidirectional modules to form a design and test link. Note that links must be bidirectional so that test data can also be back-annotated into the design environment for the purpose of resimulation, because the original design data is modified based on information gathered in testing.[1] If numerous

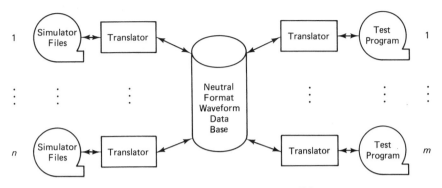

Figure 6.13 Bidirectional design and test links.

simulators and testers are used by a company, this $n + m$ approach can save a tremendous amount of resources and management headaches.

Besides converting data formats/standards, the translators must also be able to analyze the waveform information that is compatible with the tester's capabilities. Other considerations in converting the data format are the differences in the computer system environment involved in linking design and test. Most of the simulation programs run on CAE workstations while the test programs run on special tester-specific computers, and the neutral format waveform data base resides on another computer. Therefore, the problems of interconnecting different types of computers in areas such as networking, protocols, and operating systems must be considered together with the problem of converting data formats/standards.

In the last decade, significant progress has been made in this area because of the industry's awareness of the problems and the gradual acceptance of standard formats such as EDIF and JEDEC for exchanging electronic data. Many products have emerged on the market to facilitate the linkage between design and test. The end result of all this is reduced time to market and better tested products with higher quality.

6.6.2 Future Trends of Design–Test Link

With the high degree of success achieved by the industry today in the area of design–test link, we expect that it will be even more successful in the near future. The key technical factor that contributes to this is modern computer science technology, including new software engineering techniques (e.g., object-oriented programming), artificial intelligence, and computation algorithms taking advantage of parallel processing. Of course, more powerful computers will always help in the computation-intensive tasks associated with testing of complex digital circuits and systems.

We foresee one area that will stand out among tools to facilitate design-to-test links: automatic test program generation. With artificial intelligence embedded in a rule-based system, a test engineer will be able to enter a set of tester-specific requirements as rules, and then the system will automatically map the stimuli/response data from the CAE environment into a test program to be run by the tester. Only with this capability can a truly fully automated design-to-test link be achieved.

References

1. Technical staff of Integrated Measurement Systems, Inc., "Verification Solutions, A Guide to Design Verification and Test." Integrated Measurement Systems, Inc., Beaverton, Oregon, 1988.
2. Turner, Richard L., and Paul Neudorfer, "Automated Testing, A Course Based on IEEE-488 Equipment." Seattle University, Seattle, Washington, 1990.
3. Gosling, William, Twenty years of ATE, *Proceedings of 1989 International Test Conference,* August 1989.
4. S-3295 VLSI Test System Specification, 70W-5122-1, Tektronix, Beaverton, Oregon, 1984.
5. Technical staff of Logic Automation Inc., "Electronic Design and Simulation Cost, Quality, & Time to Market," technical specification. Logic Automation Incorporated, Beaverton, Oregon, 1990.
6. Balme, L., A. Mignotte, J.-Y. Monari, P. Pondaven, and C. Vaucher, New testing equipment for SMT PC boards, *Proceedings of 1988 International Test Conference,* September 1988.
7. "TesterLink 1.0 User Manual." Data I/O Corporation, Redmond, Washington, 1989.
8. "Test Library Programming User Manual, GR2282 Board Test System," GenRad, Inc., Concord, Massachusetts, 1988.
9. "PLDtest Plus User Manual," Data I/O Corporation, Redmond, Washington, March 1990.
10. Chang, Herert Y., Eric Manning, and Gernot Metze, "Fault Diagnosis of Digital Systems." Krieger Publishing Company, Melbourne, Florida, 1974.
11. Reiss, Alan J., and Alan Fitch, Fault dictionaries aid VLSI PCB testing, *Electronic Test,* June 1990.
12. Hetchtman, Charles, D., In-circuit test fixture, *Proceedings of 1988 International Test Conference,* September 1988.
13. Editorial staff of Evaluation Engineering, Design-test link software, *Evaluation Engineering,* August 1989.

Chapter 7

Special Testing Topics and Conclusions

In this chapter we are concerned with special testing topics such as mixed analog and digital signal testing, mixed-mode simulation, microprocessor testing, and microprocessor simulation. These topics are important because they address testing problems that most complex digital systems must face today. They also require special testing considerations and techniques that are not necessary in testing regular digital circuits. Some of these techniques such as mixed-signal testing are still in their infancy. Thus we must treat these topics with care and will only present some significant results achieved in industry so far. But these achievements may only be the beginning of the evolution process toward a more sophisticated and efficient testing technology.

7.1 Mixed Signal Testing and Simulation

7.1.1 Introduction

Recent advancement in the semiconductor processing technology has allowed the coexistence of devices with different processes, such as CMOS and bipolar on the same chip. This gives design engineers the flexibility to design a digital circuit together with the associated analog devices on a single wafer. As a result, combinations of both digital and analog circuits started to appear in the same IC package. Most common mixed-signal devices such as high-performance graphic chips and telecommunication chips today contain analog-to-digital converters (ADC) and digital-to-analog converters (DAC) together with digital control and other analog circuits in a single package. These circuits used to be separately packaged on a board.

It is predicted that the mixed-signal ICs will account for more than one-third of the application-specific ICs (ASIC) and that they will represent a multibillion dollar market in the next few years.[1] Coupled with the trend of clock rates and device densities doubling every four years, it would not be surprising if one day a whole electronic system containing a large number of analog and digital circuits can be produced on a single chip.

Testing mixed-signal devices is a challenge for test engineers today. The testing technology is emerging to meet the demand of manufacturing requirements of mixed-signal devices, which in turn stimulates the ATE market to meet this demand. In this section we will discuss special problems associated with the testing of such devices and their solutions. We will also discuss the design-to-test link for mixed-signal devices. Similar to the relationship between functional board-level testing and simulation as we discussed in Chapter 6, mixed-mode simulation can reduce much of the difficulty encountered in stimuli generation.

7.1.2 Challenges in Mixed-Signal Testing

There is currently a large amount of resources devoted to solve the mixed-signal chip and board testing problems in the semiconductor and ATE industry. The major challenges of mixed-signal testing facing test engineers today are in three areas.

1. definition of an analog fault
2. access of analog circuit in a mixed analog and digital device
3. test stimuli generation

There are equally tough technical problems that challenge the ATE manufacturers to provide the right testers to suit this demand. These issues together with the challenges facing the mixed-signal ATE manufacturers will be discussed in a later subsection.

Fully automated testing is prevented by the lack of a clear definition of an analog fault.[2] It is quite different in the area of automated testing of digital circuits and systems, where faults are well defined, common means of generating test stimuli to stimulate these faults exist, and the design-to-test link is well established to make the whole process from design to test a rather smooth path. The key question is "what constitutes an analog fault?" Is it a degraded performance of a device from its specification? How much deviation from the stated performance level is tolerable? What are the major criteria to judge a degradation in performance of a fault? Is it considered a fault only when its effect can be observed at a primary analog output or be detectable at a digital primary output?

There is no black and white answer to these questions. Human judgment is frequently necessary in analog circuit testing to make the failure/no-failure type of decision. In analog circuits, there are many modes of failures and corresponding effects. As a result, a fault dictionary is inappropriate for diagnosis purposes.[3] A set of measurement values must be taken to verify a working component and these values are also dependent on the component's tolerances. Some analog circuits have closed feedback paths and components with no obvious input and output (e.g., a resistor), which makes fault modeling difficult. On the other hand, for digital circuit testing, a detected stuck-at fault is definitely a failure-causing problem that will raise a flag between the test and design engineer. These are some of the major reasons for the comparatively underdeveloped position in automated testing and diagnosis of mixed-signal devices. We believe that to meet this challenge the test engineer must depend on CAE tools to make analyses and obtain data in order to make the necessary failure versus no-failure decisions and incorporate these decisions into the testing process. This will make the testing process more automated by requiring less in-process human intervention. The CAE tool that plays a key role is the mixed-mode simulator to be discussed in the next subsection.

The second challenge concerns methods of accessing analog circuits in mixed-signal testing. Certainly this can be done through the digital and analog circuit interfaces. But testing these interfaces is itself a problem today. The fundamental solution lies in design for testability (DFT) itself, just as for digital circuit testing. Among the DFT techniques, the recently standardized IEEE boundary scan methodology[4] discussed in Chapter 5 is particularly applicable for mixed-signal testing. Although the analog section of this standard is not included because the definition of an analog bus is still to be defined,[2] the basic framework already exists in the standard to facilitate testing the analog circuits and their interfaces with the digital circuits. With boundary scan cells included in the digital devices that surround the analog circuits and the interfaces between the two, access into each type of circuit can be achieved much more easily than through the many layers of interfaces of analog and digital circuitry. The boundary scan method creates a partition that isolates the analog circuits from the rest of the circuitry, while still allowing testing of its interfaces independently. We believe that this is an area in which more work is needed.

The third challenge is in the area of test stimuli generation for mixed-signal testing. The major issue involves generation of stimuli for the digital portion, the analog portion, and the interface between them. Just like in digital test generation, this requires finding a way to get to a fault from a controllable point and also a way to propagate the fault effect to an observable point. The basic notion of a path for a signal starting at one point and going to some other points is preserved, as in the digital case. But in this case, a path may have a bandwidth associated with it that can be defined as the minimum bandwidth of the analog component along the path. It may also have some kind of associated noise or linearity figures. But in mixed-signal testing generation, a path involves criteria based on gain/bandwidth compatibility as well as digital sensitization states. This compounds the already very difficult problem of generating test stimuli based only on path sensitization, as in the digital test case. Some alternatives must be used and we think the answers lie in mixed-mode simulation. We will get into a more detailed discussion of mixed-mode simulation in the next subsection.

7.1.3 Mixed-Mode Simulation

In this section, we first discuss a methodology of mixed-mode analog and digital simulation using a circuit simulator and a switch-level simulator. We will then discuss how tests are generated for each analog model used in the simulation process.

A system architecture for a mixed-mode simulator incorporates two specialized simulators[5]—a circuit simulator and a switch-level simulator—in a single computing environment, as shown in Figure 7.1. The circuit simulator is used to simulate the analog portion of a chip and the switch-level simulator is used for simulating the digital portion of the chip. The steady-state input–output behavior of the digital portion is verified by the switch-level simulator, while the timing characteristics are simulated by the circuit simulator. This mixed-mode simulator is designed especially for transient analysis of an IC. Each simulator contains a driver that controls the transmission of signals between the two portions.

During simulation, the analog driver determines at each point in time whether to send an invoking signal to the digital portion of the circuit based on the state of its own portion. If so, it will then invoke the digital driver by passing the correct signal values (i.e., current or voltage levels) as inputs to the switch-level simulator. The digital driver in turn converts these current or voltage levels in-

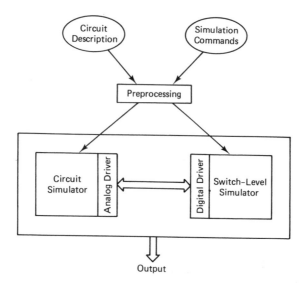

Figure 7.1 System Architecture of a Mixed-Mode Simulator.

to appropriate logic levels 0, 1, or *X*. Any signal logic values fed back to the circuit simulator are converted back to analog values by the digital driver so that they can be used to generate solutions for the analog portion of the circuit for use in the next iteration. The digital driver is then ready to accept the next input values.

The two most difficult problems in performing mixed-mode analog and digital simulation are both related to the interface between the two sides. They are conversion of an analog signal into corresponding logic values and avoidance of spurious or erroneous states resulting from misrepresentation at this interface. To attack the first problem, user-specified conversion levels must be allowed as inputs for the simulator. Otherwise, the input file containing process information is used to define default values for converting between analog and digital signal values. The second problem of spurious signals can be avoided by allowing the user to specify a settling time, and if a signal which appears at the input of the switch-level simulator is shorter than this time, then it is discarded. This acts like a filter that prevents any erroneous simulation results.

To link mixed-mode simulation[6] with mixed-signal testing, there are some key requirements in the design of the interface between the analog and digital portions of the circuit. The digital driver must be observable to facilitate the testing of the digital portion of the circuit. It must also be controllable to allow test stimuli to be shifted in directly to test the analog portion without going through the digital portion. The boundary scan type of structure suits this requirement. One alternative is to introduce an analog scan chain to observe the analog test points. It can operate in parallel with the signal being observed. But caution must be exercised to avoid the effect of capacitive loading, as well as cross talk introduced by the analog scan chain.

With boundary scan cells built into the a mixed-signal IC or board, test stimuli for digital circuits can be generated using whatever method deemed suitable (e.g., using an ATVG). However, for the interface from digital to analog circuits, arbitrary digital signals must be provided through the boundary scan chain or using a test sequence. The test stimuli for each analog circuit should be generated via a mixed-mode simulation. A simulation should be run for each injected fault, most importantly for deviations from an element's nominal values by specified amounts but also including faults for shorts and opens of elements and nets.

Some analog components (e.g., comparators) have both digital and analog inputs/outputs. In this case, the digital signal can be considered as control or conditioning signals which affect the analog path to be tested. Therefore, in test generation, sometimes both analog and digital test stimuli must be generated together to achieve a complete test of an analog component. As we stated before, the techniques for mixed-signal testing are still in their infancy. This is an area that will require much work in the near future.

7.1.4 Mixed-Signal ATE

ATE suitable for mixed-signal testing must meet major requirements such as low noise floor, open architecture, and digital signal processing (DSP) capability.[1] There are other requirements such as accuracy, low phase jitter, expendability, and throughput which are common for all ATE. We will discuss each of the major requirements individually.

The noise floor is a very serious problem for ATE that performs mixed-signal testing, because switching of digital drivers and other digital devices can produce large noise in the test head through its ground return. Since analog signal measurements in general need a high degree of accuracy, the induced noise in the ATE can ruin these measurements. Therefore noise reduction through careful ground balancing and signal distribution minimizes noise and interferences.

What type of ATE architecture best suits mixed-signal testing? The most desirable test systems are implemented with a combination of test instruments from different manufacturers. This has always been a difficult and challenging problem for the ATE industry. The solution lies in an open architecture through a standard interface bus to allow users to attach a variety of analog measurement instruments. One of such standards is the IEEE 488 Standard,[7] or the so-called general purpose interface bus (GPIB). It describes a parallel conductor digital communication bus by which a variety of instruments[8] such as power supplies and multimeters can be interconnected. Frequency response measurements, DC measurements, temperature characterization, timing measurements, and digital logic tests are all possible with an IEEE-488 system.

DSP techniques are widely used today to generate very accurate analog test stimuli and to perform accurate measuring functions. The foundation of DSP testing is to use a DAC with proper digital pattern generation to create analog waveforms and an ADC with proper digital signal processing to measure analog waveforms. It can produce fundamental and harmonic amplitudes, signal-to-noise ratios, and total harmonic distortions for various signal waveforms. DSP also allows characterization of a filter in a single step by generating all the necessary frequency components. However, DSP-based mixed-signal testing places constraints on the timing generation subsystems of the tester. These constraints are sometimes beyond the capability of the tester's digital subsystem. Therefore, the digital subsystem testing may be done at one clock rate while the DSP testing is performed at another rate. The resulting problem is the under sampling of the analog signal and associated distortion. Various methods exist to achieve synchronization between these two subsystems.[9]

The system architecture of the Integrated Measurement Systems, Inc. (IMS) mixed-signal system is shown in Figure 7.2. This system allows mixed-signal[10] prototype verification, device characterization, and production testing. It has an analog instrumentation subsystem, a digital instrumentation subsystem, and an

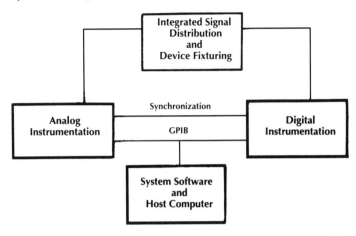

Figure 7.2 Architecture of IMS Mixed Signal Testing System (Courtesy of Integrated Measurement
Systems, Inc.).

integrated signal distribution and fixturing subsystem. Synchronization of the
system is obtained by triggering analog measurements with high-speed digital
channels.

The system has an open architecture which allows the user to incorporate
GPIB analog instruments to meet specific test requirements. It measures noise
levels, bandwidth, propagation delays, and signal interactions in detail. A key
component of this system is its low-noise analog signal distribution system to
ensure accurate measurements. Analog measurements are sequenced by placing
measurement commands at any point in the digital test pattern. Advanced dynamic tests for mixed-signal devices are achieved through test sequence control
with real-time synchronization between the two subsystems.

7.2 Microprocessor Testing and Simulation

7.2.1 Introduction

The use of microprocessors has been rapidly expanding in a wide variety of
applications. Many of these applications require highly reliable operation. This
has given rise to an acute need for efficient, thorough, and cost-effective testing
methodologies to detect faults in microprocessors. The testing problems are
made more difficult by the increasing complexity and density as well as new
architectures used by microprocessors. Gate densities that permit the integration
of an entire microprocessor on a single chip have been reached in gallium arse-
nide (GaAs) technology, and some of these chips have implemented the new
reduced instruction set computers (RISC),[11] which have unique architecture as
well as much higher operating speed.

A microprocessor's users always demand high reliability of the system that is to be incorporated into their product, and it has to be thoroughly tested beforehand. The conventional ad hoc approaches to testing microprocessors, such as testing each instruction for many operandi or exercising various modules (e.g., ALUs and registers), are not very useful because they are not based on a general microprocessor model that is applicable to systems with different architectures. Moreover they do not consider specific fault models. Therefore, it is difficult to know what faults can or can not be tested.[12]

The classic methods of generating tests, as discussed in Chapters 2 and 3 for digital circuits, are also inadequate for testing microprocessors. This is due to several factors unique to microprocessors: high gate counts in the range of hundreds of thousands, physical failures not able to be modeled by stuck-at faults, and lack of knowledge of internal design details. Combined with the fact that most microprocessors are not designed to be easily testable, this makes it necessary to consider different testing strategies and approaches to meet the associated challenge.

In this section, we discuss a general microprocessor testing strategy and then a microprocessor simulation method that is closely related to the testing strategy.

7.2.2 Microprocessor Testing Strategy

A test strategy for a 32-bit microprocessor with high gate density and large I/O pin count is discussed herein. The strategy involves a test methodology based on the simulation results[13] of the microprocessor module which consists of five functional submodules: the CPU, microbus interface, module bus controller, status latch and decoder, and the memory management unit. It uses simulation and design verification results for the module as test stimuli for module testing. The methodology is made up of certification tests of contact, DC parametric, AC parametric, and functional tests of each submodule as shown in Figure 7.3. The vector translator extracts cycle-by-cycle values from simulation vectors and maps each signal from the simulator output to the corresponding pin on the tester. This method of extracting test vectors from simulation and design verification processes greatly facilitates the testing of a complex module such as a 32-bit microprocessor. Full diagnostic tests of opens, shorts, loopback, and stuck-at faults are run if a submodule fails a certification test.

The first step in the certification test is to verify the connectivity and good contact between the module under test and the test fixture. For functional tests, the major purpose is to detect stuck-at faults and to ensure the structural integrity of each submodule tested. A set of tests for pin faults, CPU, and I/O requirements was chosen from functional verification tests which must exercise the module under all conditions. These conditions must include bus read and write operations, interrupts, and direct memory access. For DC measurements, emphasis is placed on stimulating the worst-case conditions for some parameters

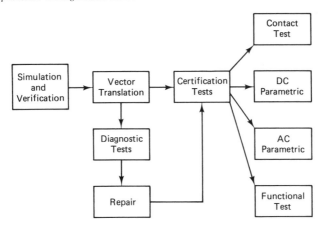

Figure 7.3 A Microprocessor Test Methodology.

such as static and dynamic I_{dd} and input leakage. For AC characterization, three types of measurements are performed: propagation delays for module outputs, setup and hold time for module inputs, and the time delay between a module's I/O signal going into tristate and a module's input signal.

The testing strategy has resulted in the establishment of a high-volume production test facility through which each microprocessor module delivered to a customer is diagnosed and fully certified before shipment. The use of software models to aid the test-vector generation process not only greatly reduces the test development time but proves to be the key reason for the success of this strategy for testing complex microprocessor modules.

7.2.3 Microprocessor Simulation Method

Microprocessor design engineers today are finding that simulation is a must for the new generation of systems such as RISC.[14] Two particular factors drive this need: fast clock rates between 25 to 40 MHz in these computers and high pin count packaging which requires fine line and multilayer board. These and other factors make breadboard prototyping techniques impractical. The extensive and expensive prototyping can easily be reduced by performing logic simulation. If we look at the life cycle of a microprocessor-based product, we see a series of design-and-try steps. In each step, verifying the previous effort is always necessary before moving on. A microprocessor module also has a design cycle that divides into software and hardware tracks. This complicates the design process because software verification depends on the availability of hardware. A hardware prototype must first be produced to verify the hardware design. The results of that process are fed back into design. When hardware is verified, hardware/software integration follows most often on a prototype hardware of the micro-

processor. Consequently, the process of building, debugging, and revising pro-
totypes is a major contributor to expense during and preceding the hardware/
software integration phase. Only when integration is completed can software
verification proceed, followed by full system verification and production.

Although simulation can reduce this expense, it had not been a widespread
practice until recently, because simulation models for microprocessors were rare.
The situation has changed in the last few years since software models are becoming available for many microprocessors.[15] Actually there are two kinds of microprocessor models: (1) hardware verification models to be used when the goal of
simulation is to verify hardware and generate test vectors and (2) full functional
models to be used when the goal is to troubleshoot software or to aid product
hardware/software integration.

As shown in Figure 7.4, when logic simulation is used, building the first
prototype can be postponed until after the hardware/software integration phase.
Moreover, the prototype will much more closely represent the final system and
will therefore be more effective during software verification. Note that hardware
design debugging and verification can be done while simulating with a hardware
verification model (striped block). The simulation runs performed during hard-
ware/software integration processes (shaded block) require a full functional
model. We will discuss each kind of simulation model in detail.

A hardware simulation model, sometimes called a bus cycle model, represents
the fundamental functionality of the microprocessor hardware needed by the design engineer to verify the design. These models are very useful during the design phase when only the hardware-related behavior needs to be verified. The
design engineer in these cases is interested in the interfaces of the microprocessor
with the rest of the system. The hardware verification models allow the design
engineer to verify if the wait state logic is correct, if the setup time on some pin
was violated, if an interrupt pulse width is within limits, and so on. In this kind
of model, the microprocessor is represented by a series of signals on its pins (the
microprocessor bus) which control the rest of the module. The combination of
different logic signal values on these pins define a bus cycle. The microprocessor
can be viewed as performing a steady stream of bus cycles, each new bus cycle
following right after the last one. Simulating a microprocessor and the rest of the

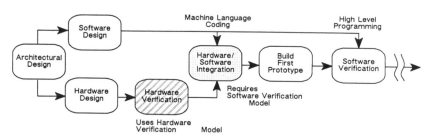

Figure 7.4 Simulation of Microprocessor Models (Courtesy of Logic Automation, Incorporated).

module in terms of bus cycles permits the design engineer to focus on the signals themselves and their timing and sequences, rather than the operand and opcodes of instructions.

Typical bus cycles are I/O data read or write, interrupt acknowledge, memory data read or write, and halt or shutdown. For instance, a simulation run can verify that the microprocessor can write to memory by first asserting the reset pin followed by two bus cycles, memory data write and memory data read. The alternative for the design engineer to verify such an operation is to write an assembly language program, debug it, and run it. This is a rather lengthy and tedious process. Another advantage of a bus cycle model is that there are only a few bus cycles for even the most complex microprocessors today, while there are hundreds of instructions.

The second kind of simulation model is the full functional model, which incorporates all the features of the bus cycle model plus simulated execution of the microprocessor's assembly language instructions. These models are required later in a microprocessor-based design because the correct execution of instructions is the essence of the system. This phase often starts at the hardware/software integration phase or earlier. The full functional model of a microprocessor can be programmed just like programming the actual microprocessor. The normal steps of programming, such as coding, debugging, linking, and loading, are similar except that the object code is loaded into simulated memory. The functional models enable the simulation of the two modules in hardware/software integration and the verification of software modules. Here the ability to execute assembly language instructions is very important.

The simulation of microprocessors based on the two kinds of models offer a way to completely develop, debug, and optimize the design of a digital system. During the early development phases when change is frequent and turn-around time is critical, hardware verification models offer fast and easy simulation as compared to the assembly language programming techniques to verify hardware designs. Later, when the hardware platform is more stable and the software must be verified, the full functional models are paramount. A combination of both models provides the most productive environment for successfully simulating a microprocessor-based design over all the phases of the design process. More important is the fact that the simulation and verification results can be ported to the testing environment to test microprocessors. The economical and technical benefit derived from this linking of design with test is invaluable.

7.3 Conclusions

In this book, we have discussed various testing problems and techniques for digital circuits and systems. We have emphasized explanations of algorithms instead of mathematical derivations of the equations involved. We have devoted pages to both established techniques and to new and emerging ones. Some areas, such as ATVG for sequential circuits, we just lightly touched due to the limitation of the size of the book. This together with some other topics can be expanded easily into a book themselves. The author wishes to express his thanks for the readers' patience in reading this book.

References

1. Jacob, Gerald, Linear/mixed signal ATE, *Evaluation Engineering,* December 1989.
2. Runyon, Stan, Plenty of activity in the mixed-signal arena, *High Performance Systems,* August 1989.
3. McKeon, Alice, and Antony Wakeling, Fault diagnosis in analog circuits using AI techniques, *Proceedings of 1989 International Test Conference,* August 1989.
4. "IEEE Standard P1149.1." Standards Committee of IEEE, New York, 1990.
5. Lee, E.S., and T-F. Fang, A mixed-mode analog-digital simulation methodology for full custom design, *Proceedings of IEEE 1988 Custom Integrated Circuits Conference,* May 1988.
6. Morrin, Thomas H., Mixed-moded simulation for time-domain fault analysis, *Proceedings of 1989 International Test Conference,* August 1989.
7. "IEEE Digital Interface for Programmable Instrumentation," IEEE Std. 488.1. Standards Committee of IEEE, New York, 1987.
8. Turner, Richard L., and Paul Neudorfer, "Automated Testing, A Course Based on IEEE-488 Equipment." Seattle University, Seattle, Washington, 1990.
9. Rosenfeld, Eric, Timing generation for DSP testing, *Proceedings of 1988 International Test Conference,* September 1988.
10. Technical staff of Integrated Measurement Systems, Inc., "IMS Mixed Signal System Specification," Integrated Measurement Systems, Inc., Beaverton, Oregon, 1990.
11. Patterson, D., and C. Sequin, RISC I: a reduced instruction set VLSI computer, *Proceedings of 8th Annual Symposium on Computer Architecture,* May 1981.
12. Thatte, Satish M., and Jacob A. Abraham, Test generation for microprocessors, *IEEE Transactions on Computers,* **C-29,** June 1980.
13. Ramaswamy, B., and P. Feder, Test strategy for a 32-bit microprocessor module with memory management, *Proceedings of 1984 International Test Conference,* October 1984.
14. Technical staff of Logic Automation, Inc., "New Techniques for Modelling Microprocessors in Logic Simulation," application note. Logic Automation, Inc., Beaverton, Oregon, 1990.
15. Technical staff of Logic Automation, Inc., "Processor Control Language and Processors, SmartModel Library Data Book," Vol. 2. Logic Automation Inc., Beaverton, Oregon, 1990.

Index

229